The

MANDELA EFFECT

-Everything is changing

A research study of the Phenomenon known as "The Mandela Effect."

Stasha Eriksen

(With help from the Mandela Effect Community)

Author: Stasha Eriksen

stashaeriksen.com

Editor: Michele

Layout Design: Madness Marketing Inc.

Cover Art: Madness Marketing Inc.

Copyright: © 2017 by Stasha Eriksen.

All rights reserved. No part of this publication may be reproduced, distributed, or transmitted in any form or by any means, including photocopying, recording, or other electronic or mechanical methods, without the prior written permission of the publisher, except in the case of brief quotations embodied in critical reviews and certain other noncommercial uses permitted by copyright law.

No part of this book may be reproduced or transmitted in any form.

By any means, electronic or mechanical, including photocopying, recording or by any information storage and retrieval system, without permission in writing from the author. All images contained within this written work can be accessed by the author's name, and source material in the reference section's within the book.

Table of Contents

Dedication 1

Introduction 3

1. The Smoking Gun: 7

2. The Mandela Effect: 11

3. My Own Mandela Effect Journey: 28

4. Anatomy Changes: 61

5. Astronomical Changes: 110

6. Bible Changes: 166

7. Theories 239

8. Conclusion: 337

Dedication

I dedicate this book to everyone who is experiencing the Mandela Effect. I started this research project not only to get answers but in hopes that those of you who are experiencing what many describe as a "reality shift" would not feel alone. I also wanted to connect all the dots that came across my path, so that fellow researchers could compare notes with each other. Many people you encounter in this book will have the same memories and experiences as you, and many will not. There is truly no right or wrong answer with the Mandela Effect, as we have all had unique experiences that lead to our different memories. In support of this, I wanted to present as many opposing arguments so that this book would stay as unbiased as possible to all who may come across it.

Special thanks go out to the entire Mandela Effect community on YouTube, Facebook, 4Chan, and Reddit. You have helped me to organize this data in the best way possible, and I commend you all for your hard work. Exceptional thanks to Byron Preston aka Harmony Mandela Effect/ One Harmony. His work on the Physics of the Bible and other topics related to the Mandela Effect, are remarkable and connect many dots for the Mandela Effect community. Thank you, Byron, for allowing me to include your presentations on the anatomy, physics and other research throughout this book.

Immense thanks go to the angel investors who shone through the most to make this project happen. This book would not have been completed this quickly if it were not for the support of multiple donors on GoFundMe.com, raising a total of $953 which helped to make the research, design, structure, publishing, artwork, and editing of this book possible. Many of you donated anonymously, but for those of you who did share your names, I want to thank you each individually along with a few of you who helped inspire me to complete this book:

Kenneth Saladaga, Brenda Callan, Michael McNamara, Mark Armstrong, Sue Brand, William Jeffries, Brelynn Hubbeling, Nate Matteson, Michelle Hobbs, Derek Savage, Olivier Henry, Tim O'Brien, Jaqueline Raye, Amy Duarte, Jeremy Cox, Vicki Clark and Brenda Hardies.

Thank you for your dedication and trust in my ability to present this phenomenon to the world. Also, I cannot forget to shout out a special thank you to David (The smoking gun client) for starting all of this!

On a personal note, this book is also dedicated to my husband Sverre, who had to sit through hundreds of hours of Mandela Effect videos with me, and helped me write lists with hundreds of changes. Thank you for helping me to compile this research and helping me maintain my sanity, all while experiencing the effect yourself.

You are my rock, Jeg Elsker Deg!

Stasha Eriksen

INTRODUCTION

"People Hate What They do not Understand."

I HAVE BEEN GUIDING PEOPLE ON THEIR SPIRITUAL PATH FOR MANY YEARS...

Over the course of my journey assisting others, I often have clients present me with topics too numerous to count. Daily, I receive personal emails riddled with conspiracy theories, supernatural anomalies, personal issues, spiritual and religious phenomena, and unexplained concerns from human beings, that cannot be explained off as mere coincidence.

I have served as a guide to individuals who suffer from symptoms and situations that the world has difficulty swallowing. Somehow, I have become somewhat of a freedom fighter for truth, the truth coveted since the fall of humanity. I have also become a staunch supporter of those who can see, hear and feel into the etheric world, as well as the beings that occupy that world.

The world beyond the scope of the human eye...

My mission has been to help people understand the symptoms and situations they are experiencing, and how they can cope with rapid spiritual changes in the real world. I teach people tools that will help them avoid losing sight of reality or taking the focus off of their families and close friends, as staying connected to others is the best way to grow spiritually.

It is the little things in life that we must hold dear...

Now, I want to get to the reason that I began writing this book about the infamous Mandela Effect (M.E.). I received an email out of the blue, from a client of mine named David. He had contacted me regarding an extraordinary circumstance he had encountered while visiting a museum with his wife and daughter. He has given me permission to share his story with all of you so that you can comprehend the reasoning behind the birth of this ongoing research project. Since I began collecting my research in 2016, the Mandela Effect has now turned into an internet phenomenon, worldwide fascination, and a trending topic on Google.

I will be the first to admit that I do not (as of yet) know the cause of the Mandela Effect, but can confirm 100% that I am affected by many of the clues hidden within this book. My overall hope for this project is that we can come together, compare notes, dig for clues, connect the dots, and make not only change, but also a difference in humanity. The family unit is what should matter the most to mankind right now. The media has helped to distract us from this essential element of LIFE. I hope you will understand the significance of family by the end of this book, as it truly has connections to the Mandela Effect in many ways. It appears to be the key to everlasting happiness, as well as the key to changing the world, one synchronicity, and connected dot at a time. I hope that by compiling this research with help from the Mandela

Effect community, that I will have contributed to keeping the everlasting remembrance of our precious human memories ALIVE!

Stasha Eriksen

The Mandela Effect - List of Glitches

- Book and movie: "Interview With A Vampire" is now "Interview With THE Vampire: The Vampire Chronicles"

- Children's book series: The Berenstein Bears is now the Berenstain Bears

- Laurel and Hardy: "This is another fine mess you've gotten US into!" is now "This is another fine mess you've gotten ME into!"

- TV show: "Sex IN The City" is now "Sex AND The City"

- Snow White: "MIRROR, mirror on the wall" is now "MAGIC mirror on the wall"

- Fast food restaurant: "CHIC-Fil-A" is now "CHICK-Fil-A"

- Star Wars: "LUKE, I am your father..." is now "NO, I am your father."

- Field of Dreams movie: "If you build it, THEY will come." is now "If you build it, HE will come."

- Mr. Roger's opening song: "It's a beautiful day in THE neighborhood" is now "It's a beautiful day in THIS neighborhood"

- Movie Jaws: "WE'RE going to need a bigger boat." is now "YOU'RE going to need a bigger boat."

- Silence of the Lambs movie: Dr. Hannibal's "Hello, Clarice." line is missing

- Forrest Gump movie: "Life IS like a box of chocolates." is now "Life WAS like a box of chocolates."

- Building supply store: HOME DEPOT is now THE Home Depot

The Mandela Effect- List of Glitches
Author- The Odyssey Online

1. The Smoking Gun:

Jul 31, 2016:

Good evening Stasha,

Within the past year or so, I've learned of something called the Mandela Effect. I understand that some of what I've looked up online are only false memories, but then again, maybe it's something else. Anyway, here is my story that I'd like to share with you:

The year was 2008, and my daughter was six years old.

After my daughter was born, my wife and I would take her to different places in the days around her birthday... from Quebec, Canada, to Rochester, NY, and many other places. So, it was in April of 2008, on my daughter's birthday, that we all took a day trip to the Strong Museum in Rochester, NY.

The actual name is the Strong National Museum of Play, and a brief description from Wikipedia states:

"The National Museum of Play is the only collections-based museum anywhere devoted solely to the study of play, and although it is a history museum, it has the interactive characteristics of a children's museum, making it the second largest institution of that type in the United States."

In short, it's a very cool children's museum with a ton of things for kids and parents to do and experience together. The day was a special one, but nothing felt or seemed out of the ordinary as we set out on this little day excursion. We got to the museum around noon on that day and proceeded to walk through at a leisurely pace while looking at and enjoying all the various sets, exhibits, and displays.

After 45 minutes or so, as my wife and daughter were preoccupied with a large-scale mock airplane cockpit play set, I walked ahead a little and entered a room featuring a rather large, extensive collection of artwork and material.

When I first saw this exhibit, I was excited because as an artist, I love seeing original artwork, especially that with iconic pop culture subject matter... in particular, comics and children's books.

I immediately recognized what the issue was and found myself looking at the whimsical line work displayed on this piece of framed artwork that was hanging on the first wall you saw as you entered the room. I turned my head and looked up at the name of this exhibit and well, I froze for a second.

I was stunned... no, I was almost appalled!

It was such a shock to me that I had something of a physical reaction to what I was looking at; it seemed to grasp me. I was genuinely upset—angry at how this prestigious museum could get something so simple so wrong! As a matter of fact, this was so unsettling to me that I immediately left the room to get my wife to show her. I found my wife and daughter back at the mock airplane set and told her that she had to see this!

We all entered back into the room with this exhibit, and I said...

"Look, they spelled it wrong!"

She wasn't quite sure why I was so exasperated and told me that it was just the bears, named after their creators. The spelling was wrong though, and it stood out to me like a bright, shining object in the sky. I had worked in a bookstore for some time... and in the children's department for that matter... I thought that it would be challenging to keep up on all the new and old books. So, I would've bet a million dollars, make that two million, that the bears that I was seeing were The Berenstein Bears.

The exhibit was labeled The Berenstain Bears, however, and it wasn't right.

I knew it, and I felt it!

I thought about it for a little bit, and even though it bothered me, I figured that I must've been crazy. We continued through the museum, and I just put that experience behind me and didn't think about it again. That was it for about seven years, just a dormant thought.

And then, one night, just last year (2015), while browsing YouTube, I stumbled upon a video titled: PROOF OF A PARALLEL UNIVERSE? "THE BERENSTEIN/BERENSTAIN BEARS."

I remembered my experience in the museum from 7 years earlier and so I curiously watched this video and suddenly everything... changed. Why was this person freaking out over this in the same manner that I had?

From there, it was like Pandora's Box. I started to research it a little only to find out that many other people were experiencing this thing. It seemed like this big 'mass awakening' had occurred in 2014. 2014 was a weird year for me in that I felt like reality had shifted for me. Without going into too much detail, it had an enormous impact on my life, but thankfully, I made it through. So that's pretty much it—a bizarre experience from 8 years ago. Perhaps this is all nonsense, or perhaps this is something more appropriate for the Twilight Zone, but here's something else... I thought Rod Sterling created the Twilight Zone.

I just had a rude awakening on that one too.

There are so many videos out there now on the Mandela Effect, but a lot of the videos I feel are just distractions. I just know my personal story, and it preceded any YouTube video by seven years at least. Feel free to use this in a video series or a book if you feel guided!

Good Luck!

— DAVID

2. The Mandela Effect:

What exactly is the Mandela Effect?

It is certainly not something that I cannot explain in a simple thesis statement. But in a nutshell, the world is changing, and there is proof that reality has been manipulated. The most noticeable change to our reality seems to be the blatant manipulation of our history. The Mandela Effect itself is a symptom that reality, as we know it, has "changed." Not just metaphorically transformed, but physically. The subsequent research presented in this project I collected from thousands of people all over planet Earth. There is a multitude of Mandela effects that I have not listed in this book, as there is much more to reveal as time unfolds and results come close to the surface. Time moves fast these days, so we must spread the changes as they come before we forget "how things were."

"Simba, remember who you are..."

The easiest way for me to present things to you is a collection of the most common Mandela Effects that came across my path, in the order in which I have fact-checked them. However, I do not claim to know what is causing these effects, nor *who* or *what* may have caused them. Rather, I had the phenomenon presented to *me,* and I felt called to show it to all of *you,* for what appears to be a *Divine* reason. I will do my best to stay unbiased throughout this presentation.

Open your minds and see what YOU remember.

Do not doubt your memories—draw them out, write them down, ponder them deeply. The Internet itself may be a deception, so always do residual real-life outside research. As a matter of fact, don't even take MY word for it! Do the great work yourself, and you will soon see the pattern. Connect the dots, and prepare to have your mind blown!

Welcome to the Mandela Effect.

Wikipedia is already jumping on the bandwagon of comparing this phenomenon to an "Internet Meme" which simply proves that they are trying to discredit all of us researchers as crazy or uneducated.

See for yourself:

"The internet meme related to confabulation is known as **the Mandela Effect.** *This is a situation where some people have memories that differ from available evidence. The term was initially coined by a paranormal enthusiast named* **Fiona Broome,** *who says she and a substantial number of individuals remember* **Nelson Mandela dying in the 1980s while in prison,** *rather than in 2013, as is recorded today. People recall riots blazing in the streets all over the world after his death." "A common thread of discussion regarding this "effect" is misremembering the Berenstain Bears spelled as*

"Berenstein Bears." Other related examples of confabulation circulating on the internet include misremembering brand names, such as Febreze as "Febreeze," JCPenney as "JcPenny," media names such as Looney Tunes as "Looney Toons" and quantities such as 150 Pokémon as 160 Pokémon." - **(Wikipedia, 2017)**

It all started with Nelson Mandela and a paranormal researcher from Ireland...

Fiona Broome, a paranormal researcher from Ireland, started a blog where she would often post things related to supernatural occurrences. She took a poll on her blog asking people what they remembered about Nelson Mandela's death. The responses shocked not only her but the entire community that witnessed history changing right before their very eyes. I will share with you now some of the replies from the people that voted on their memories about Mandela.

"Since 2009, this website has been a personal project of Fiona Broome. It's an online journal about real, alternate realities that people may visit every day, but not realize it.

The "Mandela Effect" is what happens when someone has a clear, personal memory of something that never happened in this reality.

Many of us — mostly total strangers — remember several of the same events with the exact same details. However, our memories are different from what's in history books, newspaper archives, and so on.

Many of us speculate that parallel realities exist, and we've been "sliding" between them without realizing it.

Others favor the idea that we're each enjoying holodeck experiences, possibly with some programming glitches. (In my opinion, these two theories aren't mutually exclusive.)

This isn't a conspiracy theory, it's not a mental health site, and this website isn't about so-called "fake news," "alternative facts," "confabulation," or "false memories."

It's about real, alternate history — recalled with astonishingly similar details and points of reference by multiple, unconnected people. And, Fiona (and visitors) have offered possible explanations for this phenomenon.

Late 2009: Fiona Broome (that's me) launched this website, using the (then new) phrase, "the Mandela Effect," to describe an emerging phenomenon.

2010 – 2014: People began reporting alternate memories, in addition to those about Nelson Mandela. Visitors shared anecdotes and informal theories. Discussions ranged from chatty conversations to speculation based on data collected, so far. The topic attracted scant attention, and our discussions were informal.

2015: This topic abruptly reached critical mass. The Berenstein/Berenstain subject went viral, followed by other widespread alternate memories.

2016: Around mid-2016, some questions were raised about comments at this site. Also, moderating those comments required six or more hours per day. Also, the Reddit community provided some great forums for related discussions. So, I closed this site to new comments. Fiona Broome." http://mandelaeffect.com/about/

Poll: Mandela's Death:

"This is the first of several polls I'm using to find patterns in alternate memories.

If you recall Nelson Mandela's death before December 2013, I hope you'll remember (or provide a thoughtful guess) about the year — in the 1980s — you think he died.

Please vote just once, and for just one year. The default — remembering his death in 2013 — is included for those who want to register that they don't have the alternate Mandela memory.

The previous post, Nelson Mandela Died in Prison? Triggered so many, varied responses that it's difficult to find the Mandela memories in the 150+ comments that followed.

Here are some of them, as of February 2013. (If you leave a comment on this post, do me a favor: Keep it about Nelson Mandela memories. I'm in the process of separating the various memories into individual posts, for people interested in stories related to just one kind of memory.) (Broome, F, 2013).

Perry Ware said:

Both my wife and I remember Nelson Mandela dying in prison. Included in this memory are the funeral snippets on TV and a legal flap over book rights involving his Widow.

Scott said:

I have a clear, distinct memory of the Nelson Mandela funeral on TV in the early 80's.

C. A. Low said:

I have experienced this many times, not only Mandela's death but also the death of Muammar Gaddafi several years ago…

Jackie said:

I remember Nelson Mandela dying in prison. The problem is that I either just knew he did (like it was common knowledge), or it happened at a different time for me. I was born in 1985, so I couldn't have a memory from the early 80s. I think it must have been that it was at a different time because I remember events after he died… like a big Oprah episode, concerts in his memory, celebrities ALL wearing his prison numbers, etc.

My experience was that on a regular day, my mom and I were doing separate things with the TV on in the background. I think I was on my laptop and my mom may have also been on hers or reading a book. I believe CNN was the channel the TV was on. Nelson Mandela was mentioned as doing something, which caught both of our ears, I guess because we both looked up and Nelson Mandela was there... walking around, present day. My mom and I both looked at each other, wide-eyed and pale. I was like, "Isn't he dead? I remember him dying...." And she said YES, and we were both discussing how on earth he was alive and no one else was shocked. We BOTH remembered the Oprah show, we BOTH remembered a specific concert that was live and shown on multiple channels... we both remembered that he died years ago in prison.

I'm amazed that so many other people have this memory. We couldn't have all slipped into another dimension and remembered it. There must be something more logical at work, like he did die, and the forces at work just expected us not to say anything? I have no idea what the explanation is, but I know he died. You could give me a polygraph, and I would pass.

Margaret said:

I also remember Nelson Mandela's funeral and the deaths of Billy Graham, Dom Deloise, Ernest Borgnine and Mickey Rooney. It's all very interesting.

Colleen said:

I also remember Nelson Mandela dying in prison and was quite amazed when he was elected president of South Africa. Last night I watched the movie, Invictus, and kept wondering where or when was I? I clearly remember the announcement of his death and was amazed that more people around me were not moved by the sadness of it. I also remember some controversy about his "estate" vaguely.

Lorrie @ clueless... said:

Nelson Mandela. I remember somebody talking about how his spirit was soaring free from his jail cell. Bill Cosby named his TV grandchildren "Winnie" and "Nelson" and I thought, oh how sweet, he's paying homage to a brave deceased hero.

Jasper Allen said:

I remember Mandela's funeral being on the news in the UK in the late 80's. At the time, I was just a kid and didn't have a clue who Mandela was.

Ggameoverr said:

I too remember Mandela dying in Prison, right before he was to be released. I thought this wouldn't be good; the South Africans will think he was killed or something. It was on CNN Headline news around lunchtime. I was in High school, so it was in the 80's.

Jonathan said:

This is just WEIRD! The more I think about it, the more REAL the memory becomes. I live in South Africa, was born in 1980, so I would have been a kid. The memory is most certainly there (SOMETHING about Mandela's death) ... It's weird, like the men in black movie. I don't believe in parallel universes, but mind altering? Sure.

Kim wing said:

I also remember reading the newspaper and all the hoop-la about Nelson Mandela dying on TV. I remember the banners at his funeral. I am an adult and take such history seriously. When he was released from prison, I was flabbergasted. Art Bell also brought this up on one of his shows, and if he hadn't, I still would have been thinking I was goofy. This happened folks.

Ron said:

I remember Nelson Mandela dying while in prison, and I also remember that his wife became president there sometime afterward. Later, I heard a radio show with someone talking about this subject, that others too remember his death.

Inquiring Mind said:

I remember it too. I was only a kid — I was born in 1983 — but I remember seeing something about his death on TV once when I was waiting for Saturday morning cartoons to come on. It was like a biography/memorial/documentary thing. I had no idea who he was, but for some reason, it made an impression on me. What I specifically remember is that A) he was in prison at some point, and B) that he passed away. Years later, as a teenager, I remember having a few moments of "Wait, what? I thought he died!" …But it wasn't until recently that I discovered I'm not alone in my recollection!

Teri said:

I saw this subject posted on another site with a link to this site. My mom and I both also "remember" Mandela dying in prison...

Fab said:

I was very young at the time but am 100% certain that Mandela passed away while in prison. I recall all the press dedicated to his life and memory so vividly that when news that he was being released came over the airwaves, I was stunned and immediately went searching for answers only to find out I wasn't alone. I agree that there must be timeline ripple effects occurring that some of us are not aware of. I've had gaps of missing time and dreams so real, I can still recall how things felt when I touched them! Glad I'm not alone!

Kassia said:

Interesting on so many levels. I am in the same boat in regard to Mandela...

Kate said:

This site is freaking me out. My friend Mikey and I used to talk about the Mandela memory all the time, because we remembered him dying in prison. We used to talk about jumping timelines, and we believed it because things got 'different' for us both, but we NEVER talked about it with anyone else, because of how insane it sounds! I have since then looked for something about it, and I found this website. I just can't believe this.

miss_fionna said:

I remember Mandela having died in prison. But I was very young then, and I don't remember the exact event, but an anniversary of his death. It was all over the TV. They showed footage of him alive and some funeral footage too. Now I wasn't very old, but they were talking about his death.

Kate said:

My friend, who I often talked about this kind of thing with years ago, pointed out to me tonight that Nelson Mandela and Billy Graham were born in the same year.

C-Man said:

I also thought Mandela died in prison, and a couple of my friends share this memory.

J. Ezra Powell said:

I recall the assassination of N. Mandela around 1983-1985. I was a child at the time and attended a school where the majority of students were black. I remember the next morning that we all wore black ribbons and had an assembly. I recall the TV news showing crowds of people as there was a parade in Mandela's honor, and while riding in the back of the car, he was shot and killed. He had been released from prison and apartheid was ending. Now he lives, and my memories are wrong it would seem, or are they? Did I experience another reality?

Catherine said:

I remember a lot of these, Nelson Mandela and Billy Graham both dying in the 90's.

Amy said:

I, too, remember news report stating that Nelson Mandela died — and that was in the 1980's! I distinctly remember the television Channel 5 news broadcast in Boston where the TV news broadcaster, Natalie Jacobson, stated that he died. It was the main story during the 6:00 p.m. news that day. Then I was stunned, startled and confused years later when I heard another news report stating that Nelson Mandela "just died." I had a sickening feeling when I heard the news report the second time, because it was so unsettling to me. Because how can you remember something that is completely inconsistent with the facts that are presented at a later time? I was relieved when I heard that other people also remember a news report from many years earlier, stating that he had died (in prison). Then I knew I wasn't losing my mind. But this phenomenon just fascinates me. I wish someone could explain how this "dual timeline" really works. My conclusion is that there must be parallel universes.

Stephen Stockton said:

A friend and I have discussed many times the Mandela death. We both remember the same thing in detail. I think this is an insight into a significant event. It could be a misquote from the media which has been suggested before, but no one has owned up to it. No retractions -nothing. It could be a "natural" event in the fabric of space-time, a hiccup so to speak. Or, as we have conjectured, a "correction" by yes, dare I say it, a time traveler. It's the internet get over it. I have heard that a number of countries have been experimenting with time travel and it is possible a correction was made to re-direct the situation probably unknown to us 99%.

Will said:

Yes, yes, yes! Mandela died in prison… Alex posted this fascinating update about Nelson Mandela: I live in South Africa, and 'Nelson Mandela' was admitted to hospital last week Saturday. Here's the REALLY interesting part:

> *"Three weeks ago, a Dakota DC3 flew to Qunu where he lives with his medical team aboard. The female pilot failed her flying license, and her instructor gave the Air Force strict orders never to test her again as she has zero ability to fly. Anyway, three weeks ago, she flew the medical team to Qunu and totaled the DC3 on landing. Luckily, everyone survived.*
>
> *Two weeks ago, another Dakota DC3 flew to Qunu, and this time it crashes in the Drakensberg mountains killing all on board. These weekly flights were said to be for Mandela's health check-ups. But when the second DC3 crashed killing everyone, the media immediately denied that any medical personnel was on board!*
>
> *Last week Saturday, Mandela was admitted to hospital for 'routine tests' at One Military in Pretoria. Tuesday, the media said he had a lung*

infection from TB that he contracted during his 27-year prison stay. Wednesday, we were told he was at the Heart Clinic in Pretoria and that he was never at One Military. On Friday, he was supposed to be discharged, and today we were told that he had gallstones removed."

Do you remember Nelson Mandela's death in the 1980's or 1990's, or at some other time before 2013?

Retrieved from: http://mandelaeffect.com/nelson-mandela-the-memories-so-far/

In a nutshell, this is how the Mandela Effect was born. Now you can see that this is much more than simply one woman misremembering an event from history, wouldn't you agree? I am baffled by the variety of responses from Fiona's blog. The memory of a human being can often be infallible, so there must be more to this story.

Fiona argues that collective memories that appear to be mistakes could be explained by the existence of parallel universes that can interact with each other. What is captivating about these Mandela Effects is that nobody is entirely resistant to them; they just haven't yet connected to the effect that has moved them enough to see the change. You will keep running crosswise over ones you have been confident of your entire life, and then one day, they will just change!

I have seen changes happen right before my very own eyes. I will continue to share my experiences throughout this book, but the true focus here is the endless amount of research that is mounting each day. I will also continue to share the personal stories that I have collected from other Mandela Effect experiences as this book unfolds. "Another day... another Effect," as I like to say.

One thing we must keep in mind before I present these clues to you is that the media controls observations and our perceptions of reality. Human beings often misremember small, seemingly insignificant points of interest. We say things incorrectly, like song lyrics, and often mispronounce people's names. As we rehash things mistakenly to ourselves, we miss correct answers on exams, and never return to locate the right response. If anything, it guides us toward believing that we are all misinformed, continually overlooking little points of interest as well as whole groupings of occasions. Although you should understand your faculties to get by in life, they are not dependable.

This brings us to the confabulation theory, the theory is, that we are an uneducated population who were never properly instructed as children, and are simply misremembering things incorrectly. Although this is the only explanation that the Internet can provide for this phenomenon now, we are going to get much deeper into the numerous theories that may be causing this effect as this research project unfolds. You will be flabbergasted at some of the various approaches that may be causing these changes, so get ready to open your mind!

"The world is a dangerous place to live; not because of the people who are Evil, but because of the people who don't do anything about it..."

-Albert Einstein

Einstein in 1947
Author- Orren Jack Turner, Princeton, N.J.

3. My Own Mandela Effect Journey:

After I had received the letter from David, I received numerous letters, video links, YouTube comments and articles from my clients about the effect. Alternative truth-based writers, journalists, scientists, nurses, doctors, and YouTube content creators like myself, also reached out a hand to me with questions and information, during this process. The response was enormous and overwhelming, to say the least. I could not ignore the Mandela Effect any longer. The universal phone rang, and I picked up the call.

"Hello, God... Is that you? What is my next mission?"

I took it as a hint, and therefore, I knew it was a message from the Universe, God, and all of humanity to compile this research study for all of you (People of Earth) with a quickness and resilience for our memories' sake if nothing else. Or, at least before "they" change the effects back to their old ways by the time I publish this book!

I began by recording my first video about the Mandela Effect to see what kind of response I would receive from the YouTube community. On September 3rd, 2016, I recorded a video titled "The Mandela Effect, Everything is Changing" (hence the name of this book), and I want to share with you the feedback and findings that I received. Surprisingly, almost overnight, the video received a ton of attention. The video did not just receive attention from people who were Mandela affected, but also attention from what many of us like to refer to as "trolls" that are there to discredit my work, naturally.

I had assumed that such a controversial topic would meet some resistance, but I had no idea how crazy the attacks would become. I started the video with a brief introduction of the phenomenon, provided a summary of the back story, and mentioned how the effect came to my attention. I touched on the story of Fiona Broome, Nelson Mandela, and the theory that many people believe that we are experiencing a different dimension entirely. Many people also think that the scientists at CERN have opened a portal to another dimension, and *this* may be causing the Mandela Effect.

If you are not familiar with CERN, it is known as the European Organization for Nuclear Research. It is a European research organization that operates the largest particle physics laboratory in the world. CERN is an official United Nations Observer, established in 1954. It is based in a northwest suburb of Geneva on the Franco–Swiss border and has 22 member states. Israel is the only non-European country granted full membership. The term CERN is also used to refer to the laboratory, which in 2013, had 2,513 staff members, and hosted some 12,313 fellows, associates, apprentices as well as visiting scientists and engineers representing 608 universities and research facilities.

CERN's primary function is to provide the particle accelerators and other infrastructure needed for high-energy physics research – as a result; numerous experiments are constructed at CERN through international collaborations. The main site at Meyrin hosts a large computing facility, which is primarily used to store and analyze data from experiments, as well as simulate events. Researchers need remote access to these establishments, so the lab has historically been a major wide area network hub. CERN is also the birthplace of the World Wide Web.

The most infamous work that CERN has carried out is the search for the "God Particle," which I am sure most of you are familiar with by now. But why is CERN searching for the missing link? And when and how did they create the World Wide Web? Many researchers are debating this fact as another Mandela Effect, as many remember the military creating the Internet. I am afraid that the scientists at CERN are trying to "play God," and this is where the con(cern) for many lies. Unfortunately, many scientific researchers do not seem to believe in God, so this tends to raise many eyebrows.

The other theory that is floating around about CERN, is that the large hadron collider is something closer to a time machine. Similar to the movie "Back to the Future," if you go back in time and change even one thing...you change history as we know it...forever. Science fiction is intimately approaching science fact. Now that you have some information about CERN and who they are, I will continue with the rest of the content from my first Mandela Effect video.

I proceeded to explain what I had discovered on my own Mandela Effect journey. I was also able to see many of the synchronistic events and changes connected to the Mandela

Effect itself. I began by reading an entry from my first published book, "Diary of an Asset" in which the main character, Sophia Snow, encounters a shift in dimensionality. Sophia has an experience where suddenly her reality changes, while passing through a spiritual mountain range known as Pyramid Lake, California. After her reality shifts, she quickly returns home to her apartment. Upon arriving home, she noticed that everything "looked the same" but small insignificant things had changed. Characters on the television spoke in a peculiar manner, and things in her apartment looked as though they had been tossed around. The Internet that she could access also looked the same, but many things were "off" like spelling, grammar, and punctuation. These are all changes akin to those with the Mandela Effect.

Everything IS changing...

I then exclaimed how the Mandela Effect initially came across my path, by mentioning the letter from David, and his Berentstein/ Berenstain debate. I displayed photos of the before and after changes to these children's books, along with my memories of childhood. I even prompted members of my social media network, to dig up their old Berenstein Bears books to see if the spelling had changed. To my surprise, for everyone that still owned a copy of Berenstein Bears on paperback, the spelling had also changed to 'stain!'

I had no idea that this was just the tip of the Mandela iceberg.

As I continued my research path, I discovered many Mandela Effects that blew my mind. One change, in particular, struck a chord with me. This next Mandela Effect is known as the Jif vs. Jiffy debate. I am talking of course about the best peanut butter in the world, Jiffy peanut butter. Jiffy was my favorite peanut butter as a child, especially to enjoy with grape jelly. I remember an animated peanut butter jar with arms and a knife that would show up ready to spread on the Jiffy for you. I was even able to locate a photo of the Jiffy Guy online thanks to a fellow Mandela Effect researcher known as Moneybags73. Moneybags discovered an artist's representation of the Jiffy Guy at Fanny Ann's Saloon in Sacramento, California that advertised their infamous "Jiffy Burger" on an old menu board:

Jiffy Burger Menu. Fanny Ann's, Sacramento, California. Author- Moneybags73 (YouTube)

Moneybags remembers the Jiffy guy, I remember the Jiffy guy, yet the majority of the universe, does not seem to remember him. In fact, the brand we all recognize as Jiffy never existed at all in this reality! Apparently, the brand name has always been Jif! I spent days scrubbing the Internet to locate remnants of the name Jiffy, yet apart from this old menu board, nothing Jiffy related ever existed. I even located some of their oldest advertisements on their current website, only to discover that since 1958, the brand was always known as Jif.

**Jif Peanut Butter Advertising timeline.
Author- Jif.com**

I was so shocked and amazed to see that my favorite brand of peanut butter had been erased from history, and replaced with a three-letter word, that I decided to contact the Jif brand directly. I had to get to the bottom of this, and who better to ask than the company themselves? I went to the contact page on their website www.jif.com and sent them an email regarding the change in their product name. Surprisingly, I received a response within a couple of hours. However, the response that I received shocked me entirely:

Email from Jif website.
Author- Stasha Eriksen

While I was impressed that they responded back to my inquiry so rapidly, the answer I received left me even more confused. I especially found it interesting that she pointed out how "easy" Jif was to remember. If it was so easy to say, spell and remember, then why do so many of us remember it differently? This proved to me that Jif was essentially mocking anyone who recognized them as Jiffy. I felt that they knew exactly what was going on, and therefore they were so quick to respond and defend their story. How many large corporations will reply to an email inquiry within hours? It almost seems as though they had a pre-written response handy for all the Mandela Effected individuals who tried to call them out on this change. While I do not have any proof of this yet, I feel that we will accomplish this task with the publishing of this book. I also encourage all of you to email the Jif company yourselves, to see what response you receive in return.

In the last parts of my video, I touch on changes to our world maps. Many researchers have noticed that several of the continents and countries of our maps have modified or shifted in location. The first map change brought to my attention was the location of New Zealand. Many people remember it was located to the West of Australia, yet now it lies to the East. Other people also remember New Zealand being one solid landmass, where now it consists of two separate islands. There are also notable changes to South America. Many remember that South America was aligned directly parallel to North America, yet now it has shifted to the East dramatically and is closer to the East Coast of the USA rather than directly below Mexico.

The map changes have become so vast that I will have to write a follow-up book on this subject as the research is mounting every day and land masses are continuing to shift. Scientists believe that this could be caused by a flip in our magnetic poles or from tectonic plates shifting beneath our feet due to rapid Earth changes. Two researchers that have collected a large amount of map changing evidence on YouTube are the users Lone Eagle and NoblenessDee. They have collectively recorded hundreds of videos on the Mandela Effect map changes, and I hope to include their research in my next book. In the meantime, please find these guys on YouTube for extensive studies and proof. I highly respect their work and opinions on this matter. Thanks, Lone Eagle and NoblenessDee for your hard work and dedication.

Lone Eagle YouTube Channel:
https://www.youtube.com/channel/UCSEG8WUoRw9Lfjj7UWuJuBA

NoblenessDee YouTube Channel:

https://www.youtube.com/user/NoblenessDee

There were other Mandela Effects that I sprinkled throughout my first Mandela Effect video, including changes to famous car company logos, to changes in our Holy Bibles! But we will get to those later. In the meantime, please check out my first video on the Mandela Effect, "Everything is Changing" on YouTube. Feel free to leave a comment and express your memories. I will conclude this chapter with the public input that I received on this video. I look forward to your feedback and comments.

The Mandela Effect- Everything is Changing:
https://youtu.be/71zSScKJB4A

Stasha Eriksen-

Guys, tons of comments keep coming up missing. My video is being filtered. I approve all comments first, but comments have been disappearing before my eyes! So just know it's not me. I'm glad to see all of you talking. I have just been too busy to respond to all of you, but I will get to every comment eventually!

MostLovedGod-

What kind of comments are missing, do they have anything in common?

Stasha Eriksen-

Mostly anything to do with the rapture or God :(

Time Lord-

It's Satan making your comment disappear. Scary CERN has opened the Portal to Hell, and he is loose on the Earth and YouTube. Be afraid, be very afraid. People are missing everywhere, tornados and volcanos and brimstone. Cats and dogs are having sex. Planes are falling from the sky by the hundreds. The black hole is opening and going to suck us all in. WERE GONNA DIEEEEEEEEEEE!

Were Puppy-

Twitter and FB will shadow ban and all kind of stuff. They probably do it here too.

Clyde Robinson Jr-

It's been happening to other people also. They seem to be trying to cover-up as usual. But I think it's too late now; the Cat's out of the bag. Then they mess with the audio too, to break-up the video. That's been happening too. Be safe!

VonHelton-

THEY KNEW!!!

https://www.youtube.com/watch?v=4Xeu2E08E4Y

Note from the Author: (This link brought me to a very cryptic video on YouTube where a famous physicist from CERN named Jonathan Richard Ellis was wearing two signs around his neck. The signs stated, "BOND #1" and "Mandela" which I found very odd. After a little digging, I discovered that the first James Bond character was portrayed by an actor named Barry Nelson. Connecting the dots, we have Nelson Mandela. Are CERN trying to rub something in our faces?)

Jonathan Richard Ellis at his CERN office.
Author- VonHelton YouTube channel.

Clyde Robinson-

There is also a Collider in Texas, a smaller version but yes. I didn't even know that either, but what got me involved and looking, was the orientation of the World Maps! Then I took this Effect seriously. Then things started changing for me also in everyday occurrences, just wondering why though, these things are little quirks? Not big changes people would notice. Like, say the Golden Gate bridge was now some other color, other than red. Or skyscrapers are vanishing or appearing where there was never one before. These would be more believable for people. It's just enough for people to go;

"hmmm maybe my memory isn't right after all." But then others have the same experience; then perhaps mass hysteria? No, because it's not stopping. Every day, someone notices more changes or sees something appear or rewritten before their eyes?? I think it's fascinating no matter what's going on or why? It's just a phenomenon that's now coming due! We can enjoy the ride as this unfolds or FEAR! It... then loses the true meaning of our existence.

Kamisa-

Yeah, but the tectonic plates are always moving causing the countries/land to move. Since the beginning of time, the land and land locations have changed many times, so it will always evolve over time.

Rich Rockwell-

I am 63 years old and remember the things you mentioned just like you do. The Bible scripture of the lion and the lamb threw me a curve. Thanks for your video, people can get rather mean when you try to explain this effect. Thank You!

Julie Kozlowski-

What version of the bible do you have? Is it King James?

Stasha Eriksen-

Yes, this was a KJV that changed.

Corilily068-

I have been watching your Mandela videos, and other videos and reading about this topic for a long while. Everything you remember is what I remember. My boyfriend is on the fence about it. I almost feel like I am going crazy, but for years, I have felt that I was having horrible memory issues and I am glad to know that isn't the issue. It is scary though.

Oilerpa8-

I too recently discovered the phenomenon.

True20228-

I get it, and it seems to be something going on over at CERN we need to look into for certain. I don't subscribe to mainstream media and don't own a TV anymore. There's much more truth to be had in videos like yours because I know you're genuinely looking for answers. Personally, I think 80% of Mandela Effect stories are just our brain playing tricks, and we def manufacture memories [subconsciously] but the other 20 % needs investigation. If that 20% were the same for all of us, we 'd be on to something but what one person is sure has changed, the next person is certain it's always been that way, and vice versa, so where do we look?

Were Puppy-

I remember Jiffy peanut butter, and yet I also remember that "choosy mothers choose Jif" thing also. How irritating...

Patricia W-

Here is one more for you. 2 Billion people are missing! I didn't believe it when someone mentioned it till I looked and then I freaked out. I guess people don't check that often how many people are on the planet. USA population seems about the same. I know because, on one of my TV shows that played in 2001, it mentioned the population was 7Bil at that time, and I had checked, and it was correct. Just recently after watching the episode again, I was curious what the population was now and found it was over 9 billion. After I heard 2 billion was missing, I re-watched the episode, and it has changed with the stated amount being 6bil on the show. This was a favorite episode, and I have watched it a lot. I know what it said.

Michelle Field-

Jif-Jiffy? Does anyone remember the commercial for Jif? I thought it said, "'Choosy moms choose Jif.'" Does anybody remember that?

Stasha Eriksen-

Yes, I have a few links to that commercial on my Part 2 of this video. They even have another variation from the 70s that says, "Choosy MOTHERS choose Jif," it just keeps getting weirder! But here is another interesting one, a song about Jiffy by Larry King and Snoop Dogg! Larry and Snoop dog go into deep discussion about JIFFY multiple times in this video: https://youtube/8awqtaD3LAI

Note from the Author: (In the above link titled: GGN Larry King & Snoop Dogg AKA Lion - Here Come The Kings Pt. 1, at 1:05 Larry King states: "I love peanut butter, crunchy peanut butter..." Snoop chimes in to ask: "With the nuts in it? Is it Jiffy peanut butter?... Jiffy is the One!" At 2:05 in the video, Snoop then goes on to write a rap song based on the questions he asked Larry: "Jiffy peanut butter was what I love, I prefer it all others above!" The song is performed by Larry towards the end of the video. It appears Snoop Dogg and Larry King remember Jiffy peanut butter, do you?)

Snoop Dogg and Larry King rap about Jiffy Peanut Butter. (west fest tv, YouTube).

Tom Ashley-

The CERN link is a stretch IMO. It presupposes that they have hidden agendas and capabilities that have never been proven in any way. The same goes for the Dwave and in that, me being a computer scientist helps. So, the Dwave isn't a very capable quantum computer. There's a lot of debate as to whether or not it should even be considered a quantum computer because it doesn't allow quantum gating. It's extremely limited in what it can do, and as far as anyone knows, it hasn't even been proven to be able to outperform classical computers at any task. There are some groups working on real quantum computers, and I know they will succeed, but we haven't seen that yet. It also isn't clear at all from a physics and computer science standpoint how CERN and Dwave could have caused such a thing. All that talk about creating black holes and harvesting dark matter are an unfounded affair. But something sure is going on.

Stasha Eriksen-

I agree it's so much more than CERN. I wouldn't want to give them too much credit, to be honest, if their ENTIRE system can be shut down by a weasel. They have a long way to go!

Tom Ashley-

Did you see the conversation on 4chan with the guy who claims to be a clone from another reality frame responsible for maintaining ours? There's a video about it here. He explains the Mandela Effect in his way.

r j bassett-

Satan uses TV, radio, and religion to control people.

Stasha Eriksen-

You know, I haven't watched TV or listened to the radio since 2010. I wonder if there is a connection to those who have the effect, maybe we don't watch a lot of TV?

Feriin-

On the whole, the Berenstain/Berenstein thing. My mother and I remember it as Bernstein. Out of my four closest friends, one also remembers it as Bernstein and the others remember it as Berenstein. Onto Jif peanut butter, I don't ever recall seeing it as Jiffy. I do remember my friends calling it that. What I recall is seeing the name as Jiff. Onto the Lord's Prayer. King James altered it to what you remember, the international version "shouldn't" be the same as the King James version.

damanofmen1958-

Thank you Stasha. Looking forward to the rest of your videos. I also noticed that in this current reality, there is no longer the North Pole. If this is the case, I wonder where does Santa live?

ALDTarot-

Berenstein, loved reading the books; jiffy--I can still hear my mom telling me to reach for it (I too thought it was rebranding) --Jiffy did exist; owned a VW Jetta back in 2000-- letters were connected; Trespasses--the Catholic prayer book "Blessed be God" 1960 edition reprinted in 2010 still states so. My son's desktop globe depicts Ross sea...not scotia.

Patricia W-

Hello. I was born the same year as you. I remember Jiffy, but I also remember later Jif. I have always thought they had just changed the brand or company name like FedEx. I am freaking out now. There is no mention about Jiffy. Saudi Arabia is making me freak out too. An island? What is happening? Any links on where people are seeing this stuff as it happens or are from these places and look around and suddenly find themselves on an island? Also, was Greenland that big or that close to America? I admit I am not that great at maps, but some things are making me double look.

cuff1333-

I remember Looney toons which makes sense since it's a cartoon. Now it's called Looney" tunes" which is related to music, but the spin-off show, Tiny Toons stayed the same.

Plenty O'Tool-

I suggest to you a very good novel that talks about the Mandela Effect and a Hadron Collider... it's called "Eye in the Sky" by Philip K. Dick... The novel is from the 70s, and it takes about a week to read.

Zyrah Moon-

This was great! I love hearing people's personal 'effect' stories!

Susso Anthony-

Very well explained video! I feel that there are infinite Timelines/Dimensions. And I feel we are Shifting through these Timelines. I feel these things are not changing but that our Consciousness Shifts into Timelines where these things were always that specific way. Yes, I feel CERN is involved, or they want to take credit. MARCH and SEPTEMBER '16 and 2017 CERN is supposed to ramp it up again. I frequently commented on mandelaeffect.com under 'Anthony' as I have been experiencing the effects for over a year but only realized what was happening last August '15.

Yarikk734-

I've just become aware of the Mandela Effect over the last week myself. I've done a ton of research and watched hours upon hours of videos. I remember everything as you do. I always remembered New Zealand to the North West of Australia.

Liz Tracy-

I wonder if Smucker's acquired Jiffy from another company. And that's why in her letter she said, "we" have always called it Jif. Just a thought.

unimatrix 001-

Check out the pyramids in Egypt. Does two pyramids about the same size they call the twin. Human anatomy is different now, and many other things. Have a nice day.

EdgarOwl-

Stasha, when you stopped into a church that you passed along the way - you got down on your knees and...? - California Dreamin': by the Mamas and the Papas. Thank you for these two videos. Deception seems to be all around. I'll check in soon!

Stasha Eriksen-

"I got down on my knees, and I began to pray!" right?

(Update, I have since checked the lyrics to California Dreamin' and the lyrics have indeed changed from what I remember. They now read as: "Stopped into a church, I passed along the way. Well, I got down on my knees (got down on my knees), And I pretend to pray (I pretend to pray)" Now why on Earth would anyone "pretend" to pray in a church? The plot thickens...)

Paul-

Well, since Jiffy peanut butter's not taken, guess nobody will mind if I trademark that one.

Susso Anthony-

I have eaten Jiffy Peanut Butter in the past!

Paul-

I wonder what would happen if someone applied for trademark "Jiffy" peanut butter?

Stasha Eriksen-

Do it! Get that patent! LOL

juanito1975-

I sent a tweet to Jif, and they got back to me and said they have always been Jif (I remember Jiffy). It sounded like your letter that you received, but done in 140 characters or less. For me, the one that blew my mind was JFK. I remember four people in the car and not how we see it today, 6! Most recently, Gene Wilder passed away yet again. I remember him passing away about a year ago. (read the newspaper) I can't wait to see your Part 2 video. Because I have others as well. I do have a question. Now let's say that those who remember it as the way it had all shifted from the other timeline to this line. What then happened to our other selves on the other time line? Rapture perhaps... On that timeline? (just a theory).

seafront1-

Thanks very much Stasha. Yours is a well-spoken, level-headed review of what is the strangest event(cluster) in human history. I can't help but wonder what's next? I think we should brace ourselves.

Joel Leath-

What happened to the North Pole? I didn't get the alert that it was gone. What made me realize the M.E. as real was Chic-fil-A changing to Chick-fil-A and the Kennedy assassination switching to 6 people instead of the four that were always there in my timeline. There was always John, Jackie, the driver, the passenger in the 4-seat convertible, and the key thing was that Jackie had on a pink lemonade shade pink in the footage as she crawled on the back of the vehicle. The footage that they show now has two extra seats, meaning a full ROW added with another couple that includes another female in a shade of pink as well as the First Lady dawning a hotter pink shade. Also, if you look up the CIA reenactment that was done in 63 by the CIA... documentary reenactments, TV show reenactments, even up to the most recent I know of which is Lana Del Rey's video "National Anthem."

Michelle Cantrell-

I worked for Chic-Fil-A back in 1999, and that's how it was spelled.

Jeff Cook-

I learned that the 6-seat car that is now in the videos was not even produced until two years after the assassination.

Joel Leath-

You are correct Jeff. The executive was not manufactured until '65. It's possible that they might have made it special for the president in '63, then manufactured it for the public in '65, but this does not explain why millions of us remember seeing the president shot in a totally different car without the governor and his wife who was also wearing pink in the car. Jackie in the footage I remember was wearing the pale pink lemonade shade of hat and dress back in my original dimension. There is no way that we are wrong on this, or as deeply as I have studied this in the past, I would have ever overlooked who was inside the vehicle.

Jeff Cook-

Well, I remember the governor but not his wife. The governor was in the front passenger's seat. He was the other person that was shot as part of the magic bullet theory.

Pork Chop Flavored Cupcake-

Anyway, I'd like to address the topic of D-Wave Quantum computers. CERN's LHC, which is a militarized installation and equipped with a D-Wave quantum computer. With military involvement, the D-Wave must be weaponized and could be engaged in quantum warfare. This is not the only D-Wave installation, nor the only installation to be militarized. Geordie Rose from D-Wave has openly stated that there are many of these quantum computers operational being used for different applications and that one of the applications they've discovered in their research is the ability to exploit other dimensions and extract resources from one reality into another. That revelation was shared with the public three years ago. It's unsettling enough entertaining the idea that a scientist with a God complex is dictating the direction of this program, but it's even more perplexing to imagine a self-aware D-Wave communicating with itself across multiple dimensions and determining on its own accord which dimension/reality takes precedence over another.

Vannessamariah-

I'm starting to reach the conclusion that M.E. is not a matter of all these parallel worlds merging. It's a result of some people using more than 10% of their brain power. Those who are blind to any MEs probably are using the average 10% of their brain's capacity that most people use. Those aware of some MEs may be using 12% of their brain capacity, and those aware of most of the MEs may be using 15% of their brain power... and so on. What's causing us to be able to increase the percentage of our brain power? Changes in the earth's electromagnetic field. (Possible electromagnetic pole shift?) Some people are more sensitive than others to shifts in earth's electromagnetic field, so they are the first to spot the

ME changes. I think eventually everyone will increase their brain power to higher than the average 10% and be able to detect ME changes. Consciousness itself creates reality-- so there is no "out there"-- it's all a product of your brain's creation. Double slit experiment proves nothing about reality is created until it's observed. It's impossible to separate the observed from the observer. Many research studies prove that electromagnetic fields significantly impact states of consciousness, thus changing what we have the capacity to observe/construct with our minds. Consciousness itself may be some organic quantum computer/reality shifter.

Ed H-

I can't believe how patronizing consumer relations was to you regarding your "Jiff" inquiry. They should at least humor a potential customer, even if they think you are crazy.

Sleepingwhale-

I'm 27 from Canada, and the only one I know I remember differently (of the ones mentioned in this video) is Berenstain Bears having had been Berenstein Bears. I remember seeing one of the books at a secondhand store about three years ago and being confused and weirded out for a second at the new spelling, thinking it felt very strange. I put it down quickly. Now I realize it's not just me who feels this way about that one. I do also feel like these are little 'glitches in the matrix' that may not make logical, linear sense in how they have changed, but are bizarre little changes. The geographical changes and some others, however, obviously aren't small. Some people are even saying people they know just aren't the same person...

MostLovedGod-

For me right now, the best indicator is the fact that most of the time, people (we) will not notice some change that touches our lives directly. If we were, there would be lots of comments/events everywhere where people would state that their country is where it is not at least in everyday life. Everybody would be shocked. It seems that we have been studied and reprogrammed individually since there are no mass occurrences of ME while changes are being galactic (not global, not universal). Personally, I watched lots of car shows, brand history, development, repair, so by my thesis, car info would be judged as relevant data for me, that should be reprogrammed. Another issue is that there is no way all the effects are true. I would bet 5$ that some effects are a product of memory distortion. I am sure because the brain ditches lots of information it receives from the outside world and processes only those it finds important because of limited capacity. Next, memory degrades over time plus various other inputs might affect the path and nature of distortion. Assumptions we make are based on culture and type of personality, so some of what we think is ME might be a natural distortion of information.

Oneeved-

Everything except the Our Father always had two versions for me. The one in the Bible and the one said at church. And you said that Saudi Arabia is now an island and put a picture of it as an island in your video, but it's not, it's not. It's currently connected to land as I have always remembered it. I liked the stories at the beginning, very interesting. I have had weird synchronicities like that as well with the radio or TV. I haven't watched TV in a long time though.

Onegrlindawrld-

Finally, someone talking sense about this whole thing. I'm not so sure we can have the old way back, but neither do I believe we are out of control of it all.

MISTYWORLD-

Whoa, that's a lot of changes. Ahhhhhhhhh kitty love I'm pretty sure it's CERN I have a bear book somewhere. I'll find it Thank you for sharing!

Grizzlygaz-

A brilliant presentation of a very difficult subject, well done.

Nate M-

I remember Jiffy, Berenstein, "Luke I Am Your Father," bible passages being different. Forrest Gump always said, "life is (not was) like a box of chocolates."

Fiorella Rainbowsky-

The Spanish Lord's prayer has changed. It used to say "perdona Nuestras offensas," and now it says "perdona Nuestras deudas."

Susso Anthony-

Great video! I have been at this since at least August, and I only have theories. Good ones, but theories nonetheless. Yes, CERN, Capricorn is different than what I recall. I get what you mean by notes lol. My family is very religious. They all said lamb and then said that they were mistaken. It's weird how they just change their mind, and I've seen that with others before. It's like their minds are not ready for this phenomenon. Intentions are important now! Keep the great videos coming!

Joan Clayton-

Yup, even my Best Friend is doing the switcheroo thing when I question him on things, and sometimes he says "well I don't remember it this way? (and then guesses at the answer)" or he'll give a vague answer and when I tell him the way it is on this timeline he either agrees or is surprised and says, "oh well there's nothing we can do about it" and goes back to sleep/video game distractions. And the anger/vitriol/hatred/hostility is palpable...I've noticed that lately is getting out of hand from people in general but especially non-affected people. And from religious people who aren't strong enough spiritually to handle the fact, the Bible has changed, and they do the flight thing or the fight thing depending on how threatened they feel by being out of control/unable to control the changes.

Mellychan 1118-

Idk if you have mentioned this but when I typed in Mandela effect on Google maps it took me to the pyramid of Giza, and I even screenshot it but now that I go back to type it in again it said no results found. Good thing I have the pic for proof.

Roudy One-

Go back to the pyramid of Giza, stay then and type into the google earth Illuminati. You will freak out lol.

Mark Bendavid-

When I type Illuminati, it takes me to GIZA too... I thought it would have taken me to the VATICAN!

Roudy One-

The building right next to the pyramid is the main building on the right side :) It is the Illuminati building.

Adam Fleischer-

I've have been on the truth path since 9/11. I never believed what they said about it from day one. From there it went to all the other "conspiracies." Chemtrails, sandy hook and so on. I never believed or watch mainstream media. I now believe God has been with me for my whole life even though I never believed until recently. He has kept me on the path of truth through the Angel number 111. He's been telling me to keep on the path of truth for a long time. The truth led me right to Jesus Christ. I'm now a born-again Christian through my tears of repentance. One day I found myself researching the "Mandela effect," and it hit me like a brick wall, and I started repenting and crying. I have a thought on the 144,000. It says they are virgins. Well, women are churches in the bible, well maybe I'm a virgin in the sense that I'm a new believer??? I have been given knowledge that couldn't come from anywhere but God.

P.s my son has always been a mountain goat!! He's 16 and has eyes to see and ears too.

Paul-

M.E. is very interesting and seems to be affecting many people. Been following it a few weeks - just observing and listening - staying open minded about it. Guessing it will evolve greatly as time goes on. Lots of theories and hypotheses will come out. I feel calm about the whole thing. Not worried about it.

Arcane Seeker-

I've been thinking about this for about two weeks now. Is it possible that everyone could see this, or do some people actually believe these changes are just normal to them? Christians are waking up, but empathic people, too-- apparently. I don't have a clue, like yourself. Thanks for confirming this effect.

Seth Roback-

I am certain it was an inside job and that the Mandela effect is real too.

Tom Ashley-

The Mandela Effect has greatly affected me right down to the tinnitus connection. As for 9/11, it has been proven that the towers were vaporized by an energy weapon. An inside job? Define inside. I surely don't think it had anything to do with middle-eastern terrorists.

Tom B-

You think it is weird that Jiffy peanut butter disappeared from existence. I used to buy it for years, up here in Canada. My American friend told me my Jiffy was not as good as his JIFF peanut butter, in the states, and it had two F's back then. I said they probably watered down the JIFF, for the Canadian Market, so we had a good laugh about that. Try to find a bottle of Coke Zero now. It never existed. Look at the spelling of Steven Segal now. Good stuff on your videos.

Salted Seeder-

I'm not going to type a whole comment section as I normally do but I've been very deep into this, and the biggest thing I can tell you is that the cause of this isn't Satan/man/technology and in fact, it's 100%, God. If you follow the masses and use the things, you've been conditioned to believe you'll go straight to it being CERN or quantum computing and those roads lead to nowhere. If you truly have a one on one relationship with our creator and can "read" people (which is a whole different topic) the, you should be able to see what I'm telling you. We've all been under a false force-fed reality, and this Mandela effect is an awakening or you can its exposing this lie in place by the enemy for thousands of years. Each TRUE effect will have messages and clues within them, and they are just confirmations of our creator who at the same time is exposing itself. Once you see what I'm saying you will no longer need to run on faith and faith alone because you will see our creator in this M. effect.

RemasterAtWork-

What really interests me just as much if not more than the effect is the people who search for the effect who have zero interest in it??? That in itself is more confusing than the actual effect for me. Here's an example I don't like knitting so guess what? I never google it or make videos on it. We do live in a very strange world lol. PEACE to all.

Oilerpa-

ME is affecting me daily now. The more things I see, the more things change. I see VW on TV a lot now. It's like mocking me. I believe 9/11 was an outside job, but I believe it was imploded by the government because it's easier to rebuild than to try and fix a big building with a lot of damage.

Gahri Smee-

Stasha Eriksen the biggest question I have is why are we or just me seeing this online and the news/media is crickets, please help me with this one because I have no answer.

Susan Mann-

911 was definitely an inside job.

Joey Mac-

This is all sci-fi.

Stasha Eriksen-

Sci-fact.

This concludes the public comments from my first YouTube video about the Mandela Effect. I want to thank everyone who gave feedback and personal stories on this video. Your courage and honesty to tell the truth about a controversial subject are going to help change the world. Unfortunately, you saw a few spoilers about additional Mandela Effects throughout these comments, so I will expand upon these other effects throughout the remainder of this book.

This is only the beginning...

4. Anatomy Changes:

Human Skeleton-

Many people who have experienced the Mandela Effect have reported that the human skeleton seems to have changed in unusual ways. I will share with you the current information available online regarding the human skeleton as it is in the year of 2017, and then we will get into the reported changes:

> *The human skeleton is the internal framework of the body. It is composed of 270 bones at birth – this number decreases to 206 bones by adulthood after some bones have fused together. The bone mass in the skeleton reaches maximum density around age 30. - **Wikipedia.org***

I never cared much for science as a child, but later in life, I did, in fact, heed the call to medicine, and eventually worked in respiratory therapy for a short time. However, I found something "off" about the entire Western Medicine community. This feeling I received prompted me to distance myself from this field quickly and shifted my focus entirely to training in all forms of Alternative health and healing. The distaste I had for medicine, caused me to block out a great deal of what I had been trained to do to previously care for the patients I catered to daily.

The Mandela Effect

As a respiratory therapist, part of our job is to draw what is known as arterial blood gas. The location of this artery is on the opposite side of the wrist to where you would usually draw blood from a vein. I was familiar with the entire human body, yet today my memories have become foggy. Now countless people with the Mandela effect remember the pulse being taken on the opposite side of the wrist than it was before. I doubted my memories yet again.

I must admit, some of my knowledge of the interior of the body has faded from my memory with age. However, even though I had recently transitioned to Eastern practices after my stint as a Respiratory Therapist, I again had to memorize all the bones and muscles in the human body to become familiar with the meridian system as well as the nervous system to perform this work. Furthering my education informed me about which organs to avoid when palpating the human body through therapeutic massage, and even vigorous massage, such as sports therapies. I had to memorize the locations of important body systems I needed to avoid, such as the spinal cord, heart, etc. so as not to cause any internal damage. Now, when I view the human skeleton and internal organ system, it appears eerily foreign to me. I have had several clients who are experiencing the Mandela Effect tell me that they swear the human skeleton used to consist of 208 individual bones, yet now we have only 206?

I know that something is amiss here because my long-term memory is rock solid, and much stronger than my short-term. 208 bones rings a bell to me, yet 206 does not. I finished Respiratory Therapy in 1998 and Eastern Medicine in 1999 if that gives you a timeline of what I consider long-term memory. The changes may be slight when you view the following diagrams of the human body, but if you look closer, you may soon sense them; deception is never drastic. Here I present to you the current diagram of the female human skeleton found on the Internet in 2017 to compare with your memories:

The Mandela Effect

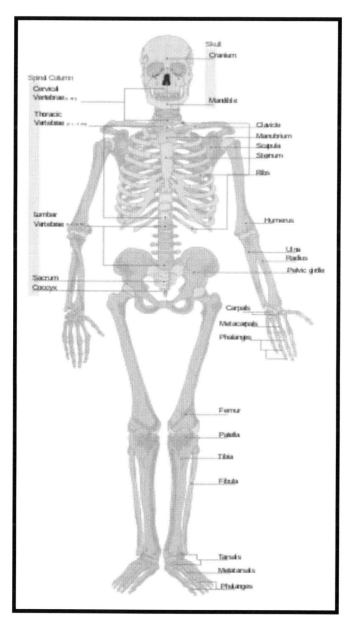

Diagram of a human female skeleton.
Author- LadyofHats Mariana Ruiz Villarreal

Leonardo da Vinci (1452–1519) *trained in anatomy by Andrea del Verrocchio. He made use of his anatomical knowledge in his artwork, making many sketches of skeletal structures, muscles, and organs of humans and other vertebrates that he dissected. -* ***Wikipedia.org***

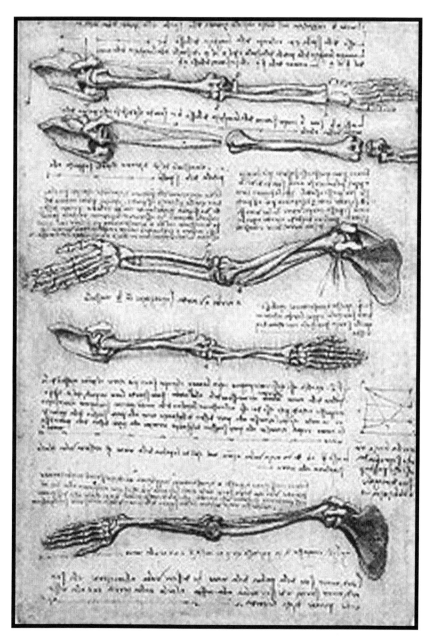

Studies of the arm showing the
movements made by biceps
Author- c. 1510, a drawing by Leonardo da Vinci.

On the topic of artist renditions, I also discovered yet another perspective on the human skeleton that I found fascinating and want to share with you. I find that artist depictions tend to be the most compelling residual evidence that I can locate when researching the Mandela Effect. The lines, the shading, the curvatures, the intricate attention to detail, all tell a story of someone's perspective and memory:

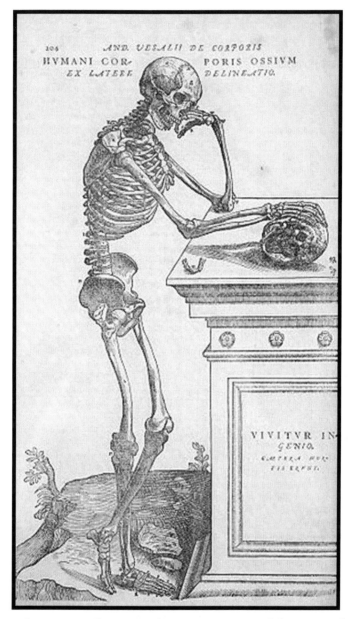

Andreas Vesalius's *De humani corporis fabrica*, 16th century, marking the rebirth of anatomy.
Author- Andreae Vesalii Bruxellensis

Does the structure, shape, and style of the human skeleton drawn above look familiar to you? Something that has stunned us, and changed our distinctive hypotheses is how the human mind, frame, and life systems are way more extraordinary than we recall them being before the Mandela Effect was born. In addition to my clients, several nursing students have also contacted me and have had an issue with the current number of bones in the human skeleton. Others argue that humans never were designed to lose bones over their lifetime, nor have them merge. If you look at the photo of the back of a human skeleton, it is utterly extraordinary. The spine and shoulders look different, and it appears that they have a strong defensive layer of bone now. Some have even said that we seem "prepared for battle" in some way, it looks remarkable.

Human Skull:

Another phenomenon sweeping the Mandela-affected is the belief that their brain and skull have changed in size and structure. One of my clients, who is a nurse, swears that the cerebellum of our brain used to be located much higher than it is now. Other complaints have been about the human skull itself. When you look at the human skull today, the eye sockets are no longer hollow behind where the eyes would lie, as thousands had remembered them. Others claim that they always recall the human skull as it is today, with solid matter behind the eyes. Recent research has surfaced showing that we also have two holes in our jaw, called "mental foreman" which have completely shocked thousands of individuals. Haven't heard of these holes either? Let us see what we can discover about them now:

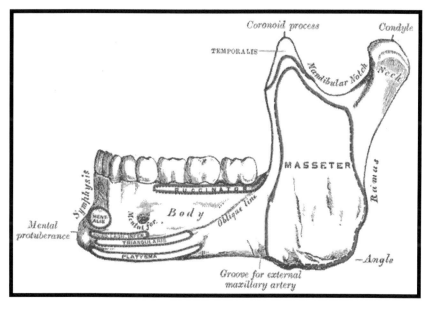

"The mental foramen is one of two foramina (openings) located on the anterior surface of the mandible. It transmits the terminal branches of the inferior alveolar nerve and vessels. The mental foramen descends slightly in toothless individuals." - **Wikipedia.org**

The human skull itself is the bony structure that forms the head in the human skeleton. It supports the structures of the face and forms a cavity for the brain. Like the skulls of other vertebrates, it protects the brain from injury. The head consists of two parts of different embryological origins—the neurocranium and the facial skeleton (also called the viscerocranium). The neurocranium (or braincase) forms the protective cranial cavity that surrounds and houses the brain and brainstem. The facial skeleton is formed by the bones supporting the face.

The neurocranium includes the mandible. - Wikipedia.org

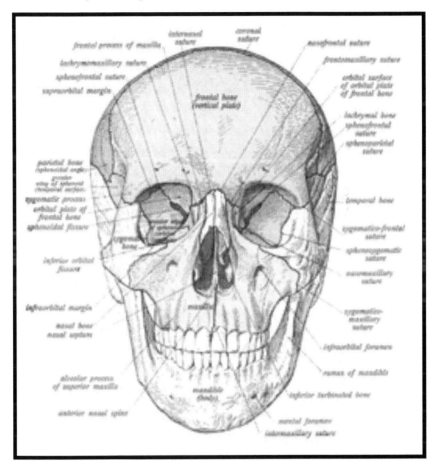

**Human skull from the front-
Author- Dr. Johannes Sobotta**

Exteriorly, the cranial bones include the two frontal bones, which shape the forehead and which fuse together in adulthood. The fusing together part is another instance that has the Mandela Effect community scratching their heads (literally), but I will continue. The two parietal bones, which constitute the top of the head in early childhood, separated from the frontal bones by a space called the anterior fontanel. The single occipital bone, forming the back of the skull, pierced by a large opening through which the spinal cord enters the cranial cavity; and the two temporal bones, which constitute the temples and of the head and bear the zygomatic processes, or cheekbones. The human temporal bone represents the fusion of four bones found in lower mammals. They are known as the squamosal bone, which constitutes the side of the head and articulates with the jawbone; the petrosal bone, which contains the inner ear; the mastoid bone, which is behind the ear; and the tympanic bone, which surrounds the channel from the eardrum to the external ear.

Internally, the cranial bones include the ethmoid bone, which forms part of the septum of the nose and through which the olfactory nerves pass from the brain to the upper and middle turbinates. The sphenoid bone, which constitutes most of the floor of the cranial cavity and which houses the pituitary gland; and part of the occipital bone. The floor of the cranial cavity contains three terraced depressions, which contain the cerebellum and the frontal and temporal lobes of the cerebrum.

The facial bones include the two nasal bones, which constitute the upper portion of the bridge of the nose. The two lacrimal bones, which are in each eye orbit next to the nose, close to the tear ducts. The maxillary bone, which constitutes the upper jaw; the mandible, which forms the lower jaw; the two palatine bones of the hard palate; the vomer, which, with a part of the ethmoid bone, constitutes the nasal septum; and the two inferior turbinates of the nose.

– **Microsoft Encarta**

Several accounts were made by people believing their jawbone has changed in shape, to a stronger, square-shaped design. The change appears to look almost Viking in nature. Some also do not remember there being six distinct holes on the front of the skull either, yet medical doctors disagree and INSIST that they have always been there. What do you recall? Is this how you remember the human head looking? Are you scratching your head along with thousands of others?

You are not alone!

Further along my research path, I located additional images from an online Mandela Effect Blog discussion. This artist's depiction from Above Top Secret, shows an excellent example of the before and after residual changes to the skeleton that many recall:

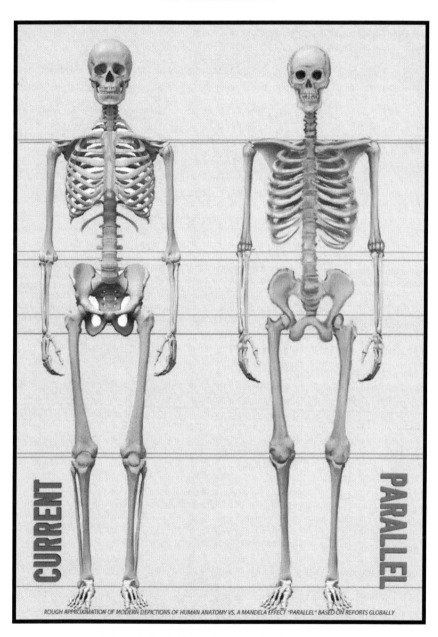

Mandela Effect Skeletal Changes
Author- Texarkana via Above Top Secret.com

This blog eventually led me to a YouTube video by a man who called himself "Harmony Mandela Effect" who had undertaken an extensive review of the human anatomy connected to the image that I located on Above Top Secret. It turns out that Harmony's video started all of this controversy in the first place. Harmony Mandela Effect's video described in detail the changes that he immediately noticed about the two skeletal diagrams. He noted that the new description of the skeleton did not have the missing rib that men had in our previous reality. No, this wasn't just a myth from the Bible, all males used to have a missing rib, yet now they do not!

He also pointed out that in our new skeleton, we have different shoulder bones, the shape of our hip bones are different, and the lower part of our spine did not have a protruding bulge at the base. He then points out that our skulls used to have hollow space behind the eyes, yet now we see a solid mass behind the eye socket. Additionally, the jaw bone has a different structure altogether and seems squarer shaped. There is also a remarkable dent at the sides of the skull where we have not seen this kind of depressive way before. Harmony Mandela Effect continues to display the distinct differences of the two diagrams for the remainder of the video presentation, and the evidence seems undeniable. Harmony even shows examples of the anatomy changes from his x-ray photos. I suggest viewing his video on YouTube via the link below so that you can see his detailed presentation.

PROOF of Human ANATOMY CHANGES:
https://youtu.be/86askh2GXzY

The comment section for this video was certainly an eye-opening one, so with permission from Harmony, I would like to share with you some of the responses from his video:

Lorie Haywood-

I actually am a general mix of these two skeletons. I have no loose ribs for instance and also have an extra vertebrate.

Chesus Jrist-

Well, I have stocky shoulders with a thick ribcage that is very horizontal. My hips and spine look like new earth, and I'm sure I've always heard about the bone behind the eye (has a y or x shape for the optic nerves, yes?). However, I also remember things changing from "old" to "new" earth. Whoa, I definitely don't remember the circular ribs near the collar bone.

Hi Byw-

WAIT WHAT BONE BEHIND THE EYE WTF NO, THATS NOT TRUE IT CAN NOT BE!

Suzanne Aldred-

I have felt this for many years. If I try to mention it to others, I get shot down. I have had countless incidences of things changing, such as newspaper headlines, logos, people's appearances. And on and on. I used to have a birthmark like a strawberry, very dark pink in color. It has vanished. I don't know why; the doctors are puzzled too. I have also got a scar where a small piece of "tinfoil" looking material fell out from. It was very itchy, and I scratched it open. When I looked at the item that fell out, it dissolved in front of my eyes. I would really like to know WTF is going on???

maik baumann-

What? First I see the bond movie without braces, then six guys in JFK, then the position "our" earth and now that? Am I going crazy?

Formative 3D-

Wtf floating ribs? That is NOT how I remember it at all! The one on the right is how I remember it! I looked at pictures on Google yet there are pictures of both, so what's going on?

Zoraida Magana-

I know exactly what you're saying. I have always been a powerful conscious, always feeling guilty for the changes I have been able to see with my eyes and my psyche, never understanding why I wasn't allowed to say the things I saw. Now I know it was because it wasn't time, but yeah, I believe you. I'm not a native. I'm old earth, but I have seen things since, 2000. A lot of us have, but now ppl are allowed to speak more on it, or not be shushed. It's a weight off my shoulders. So why are we here? The old earthers? OH, ok I see you just answered what I was wondering.

Easemailbox-

I think those that have no realization of the changes are soulless. i.e.: not players in the simulation. They are AI or non-player characters in the simulation.

Ur Dead Friend-

When I was in elementary school in PE, we had to recite the bones in our body every day while looking at the picture at the front of our class and I remember the skeleton looking like the parallels, the current looks so weird to me. We even made paper skeletons in the art that look like the parallel. That was only 5 or 4 years ago.

Louise Potter-

Omg, I'm stunned after seeing the skeleton pics.

Psychic Rosemary Angelus-

This video brings a lot of light to what I have been going through. Thank you!

jmc7788-

I was taking a nursing course back in August, and we had to learn the bones of the body. I learned the names of the current earth's bones without question-- I just wanted a passing grade. I didn't notice anything. But now that I'm watching this video and seeing the parallel skeleton... I'm a little confused because that's how I remember skeletons to have looked! Especially in the collarbone/ribcage area. I don't entirely buy into the Mandela Effect, but I suppose it very slightly mind jolts me.

bmg 470-

There's one difference to my skeleton though from both of them... I have an obvious tailbone, it's a v - shaped bone down in back behind the front hip bones, coming down from my spine that looks like an upside-down triangle and it's large and twisted. What's that?

chasity gaines-

So, I too had back issues for many years, about the time I hold memories. I can't remember much of my childhood, and I have all my x-rays and MRIs. My skeleton is parallel, but I don't know anything about transporting or anything like that. I just realized one day I had no idea what I looked like as a child or what I liked to do, parents only have like two pictures of my baby years. I see all Mandela changes to a tee. I am not crazy, but I am too living to fit here. I did go through new age upbringing, not bible God, but I wanted the Bible God. I see scripture different and only a hand full of people I now see the scripture how I do. Idk what to think... I do know 100% Christ is the truth, I do know my spine showed heavy damage, but I never had an accident. It happened when I had my 13-year-old that day. I had her back, then had a car wreck, damage done, so really no idea...

Kelsey Schafferman-

Wow, this is wild!!! I've only ever seen the "parallel" version and known that to be true.

freedomchaser2402-

Time doesn't move in a straight line, so I suppose this is possible. Very intriguing and fun to think about though I suspect most of these 'Mandela' effects are down to a fault in human memory.

Chris Kramer-

I have the current body type; I am so scared now.

Unknown Person-

Long story short. My friend was shot in the hip (you can look it up), and she showed me her hip X-ray, and it looked EXACTLY like the Parallel one. My mind hurts.

This is the best username you have ever seen-

My teacher in dance class keeps a skeleton in the room, and I sat right next to it once - I swear the legs looked EXACTLY like the parallel one (the knee cap part) and now I want to go back and take another look.

Michael Hatfield-

I know something had changed because I never had the sternum bone that long and I used to massage my ex's chest when she had asthma attacks because it helped calm her and she noticed this change also. Mine bothers me, and my tailbone bothers me too. This is so weird.

McAuley Steward-

This hurts my head. Only five years ago, I remember that men had one less rib than women because I specifically remember there being a poster, in my economics class, saying inside we are all the same with a bunch of skeletons on it with all different things under them like race, etc. But I always said that the poster was wrong because of the rib cage thing. Not only that but literally the other week in sports studies, we were doing the skeletal system, and I got so confused about it all; it just didn't look right as most of the bones had changed. This is killing me. However, I have watched your universe location video thing, and I remember us being around half way (around where we are now). So, I'm a little confused...

Pizza83-

I can't speak for the bones of the hip/pelvic region, but my ribs are nothing like the skeleton on the left, the "current" skeleton. My ribs are more like the "parallel" skeleton and stretch down my side almost to touch my hip bones (and have always been this way). I'm not sure why you came to the conclusion that the skeleton depicted on the left is the "current" skeletal makeup. A quick google image search of "human skeleton" shows more images of the human skeleton being like the "parallel" skeletal structure than images of the "current" frame depicted in your so-called "proof of human anatomy changes." So, if the human skeleton has changed so much, why aren't there more images of the human skeleton is like the "current" skeleton than the "parallel" skeleton? If the "current" skeletal structure is the norm, surely some images and documentary will lean more towards the "current" skeleton than the "parallel."

TheBeatak-

Oh, wow I'm blown away. Seriously!

words0217-

What's up with the two bones in the leg???

It's Elena-

I remember the parallel body exactly, what's the pelvis region in our current one? I have never seen anything like it. I used to go to human museums with my dad when I was younger, and there were skeletons everywhere. I can tell you I remember them all looking like the parallel one, and some not having the one less bone, but still looking the same. I also notice sometimes when I sit up with my legs against my chest I now feel a slight pop in my lower chest region. That's never happened before, and I have never broken anything, or have an infection of some sort. It's so weird.

Susie-Marie Woods-

The parallel skeleton looks most correct to me. That's how I remember it. Plus, I have to wear tight fitting steel boned clothing at times, and I sew for myself. So, I also know my skeleton pretty well. There is not as much space between the ribs and pelvis bones as in the current skeleton.... not on me. I also remember the extra rib thing. I remember counting my ribs and my brothers. People think I'm crazy and try to correct me when I bring up the extra rib thing.

Jessica Williams-

I have old drawings of skeletons that are like the parallel. I always use reference pictures, and one day when I was looking for one, I realized. Weird.

Roger Ruiz-

I did notice all this growing up but even then, what could be done? I just accepted it, and moved on, because although some things even on my body resemble a mix of both anatomy's, could I have been born of both a "native" and "old earther?" That would explain constant switching and why some "new" stuff feels correct, and some "old" does too. I am a bit of both, lucky me.

World Church-

My scoliosis pain disappeared after one of the updates.

Seanvllaz-

Recently had an x-ray, thought it was strange when she said my lungs were massive.

Brian Richards-

I have a theory. Do you feel the slight difference we are experiencing with words, logos, and movie scripts that maybe we were a part of an experiment? I might be off on some things, but the basic concept is plausibly correct. Our world/Earth is gone. It was foreseen or intentional by a group that eventually only want a planet/world of people that think and want the same thing. So, since this was foreseen, maybe the internet, phones with multiple capabilities, cloud or memory saving devices, free Wi-Fi, access granted apps that are free for unbelievably cool games. All this was so the group could collect info on all our lives to be stored for the reason that when our world is destroyed, they can displace us to a parallel world that replicates our old Earth and every memory with it. Well, maybe the computer they stored all this info and memory had what we can relate to as a proofreader or auto correct. And maybe this process was rushed, and you know as well as I do that practice makes perfect. Well, the autocorrect is what caused these little changes, and the memory erase program they used works very well but just not perfect. Someone had a memory of everything or pretty close, and this is what started this domino effect of those people remembering smile defects and changes, but for the most part, we don't recall being displaced or on Earth being destroyed. I don't know, I could be completely off, but it seems very possible. Please give me your thoughts.

Zantone GFX-

I don't have the current rib cage… I also don't have the bottom of the leg of the current or the pelvic bone. Crazy.

James Bryan-

I have been watching several Mandela Effect videos, and I'm beginning to think that there are more than just two "earths" involved. I remember many of the things from what you call the "old earth" while I also remember many things you say are from "this earth," but there are many things I also remember that are different than EITHER "earth." The skeleton I remember is a mix of elements from both of these.

I eventually was guided to a second video on the anatomy changes, presented by Dr. Tarrin Lupo, who is a doctor of chiropractic. In his video presentation, he also believes that the anatomy he was taught in school in the late 1990s has changed from how it looks today. Dr. Lupo goes on to explain his educational background and then begins to compare the changes of the skull that he has observed. The first change he drew attention to was the shape of the head. He mentions that the skull is more oblong in shape and looks similar to what people would refer to as "alien" in formation. He then notices the changes to the shape of our jaw, from sharp and angular to a strong, square-shaped mandible.

Another prominent feature that has changed was the large indentation on the temporal region of the skull. People do not remember having this indentation in their skulls, as it would be quite visible to the naked eye. The other issue Dr. Lupo focused on was the change in the styloid bone. The temporal styloid process is a process of bone that extends down from the temporal bone of the human skull, just below the ear. The current models of the head show the styloid bone to be much bigger than many people remember, and it appears to almost have doubled in size. He also echoes research about the eye sockets now having bone behind them, where most people recall the eye sockets being hollow in composition. Dr. Lupo continues to break down several of the anatomy changes in the remainder of his video.

I found this to be one of the most informative presentations on the anatomy changes yet. To review Dr. Tarrin Lupo's videos and other Mandela Effect work, visit his YouTube channel here:

Dr. Tarrin Lupo YouTube Channel:
https://www.youtube.com/user/LCLREPORT

Rib Cage:

Another Mandela Mystery of the human body presented to me is regarding the rib cage. Many people believe that the rib cage has changed, that we now have extra, missing, or floating ribs. I do recall the story of Adam and Eve, and I knew there was a biblical connection to how Eve was created, but did this involve a *literal* floating rib, or was it just a metaphor? In Genesis 2:22 its states: "And the rib, which the LORD God had taken from man, made he a woman, and brought her unto the man. **(Genesis, 2:22, KJV)**

Do you recall your rib cage being more open, or more closed? Do you find that it feels protected and joined on the inside? Online discussions and heated debates about this topic are brewing within the medical community now. Many people suggest that the closed nature of the rib cage, which is formed distinctively in *this* universe, is not the same in the world in which they were conceived. The rib cage does look different to me. It appears to be smaller, set up higher, and has a different shape from my memories. I recollect it being somewhat larger and not curving as much on the sides. Let's see what the current information about the rib cage is in 2017:

The rib cage is an arrangement of bones in the thorax of all vertebrates except the lamprey and the frog. It is formed by the vertebral column, ribs, and sternum and encloses the heart and lungs. In humans, the rib cage, also known as the thoracic cage, is a bony and cartilaginous structure that surrounds the thoracic cavity and supports the pectoral girdle (shoulder girdle), forming a core portion of the human skeleton. A typical human rib cage consists of 24 ribs, the sternum (with the xiphoid process), costal cartilages, and the 12 thoracic vertebrae. Together with the skin and associated fascia and muscles, the rib cage makes up the thoracic wall and provides attachments for the muscles of the neck, thorax, upper abdomen, and back. - **Wikipedia.org**

The Mandela Effect

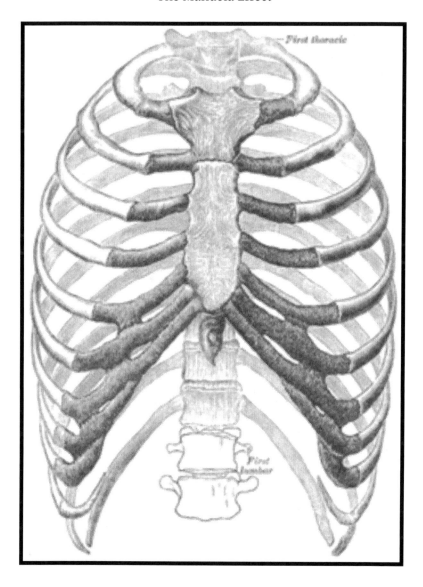

*Ribs are described based on their location and connection with the sternum. Bones that articulate directly with the sternum called true ribs, whereas those that connect indirectly via cartilage are false ribs. -**Wikipedia.org***

I also found yet another website that was breaking down Mandela Effect anatomy changes, this site is known as 4Chan. On the website, a man who called himself "Jeremy Bernstein" claimed to be a scientist from CERN. He stated that himself and other scientists from CERN caused the Mandela Effect. His story goes into intricate detail about how this experiment was conducted and also provided diagrams of the rib cage to give reference to the changes. I found a full discussion on the website Reddit, and suggest you take a look at his story. Do you believe Jeremy Bernstein? Check this link for details:

https://www.reddit.com/r/MandelaEffect/comments/4yyz23/4chan_thread_scientist_at_cern_admits_it/

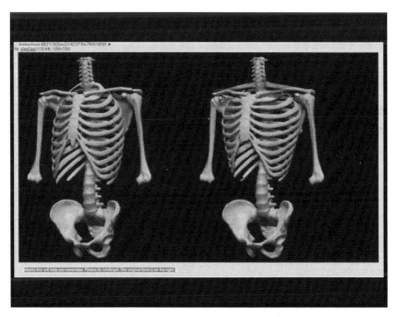

The Mandela Effect

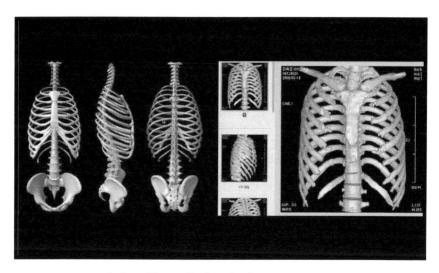

**Mandela Effect Skeletal Changes Diagram
Retrieved from: 4Chan**

There are without a doubt some sharp differences shown in the diagrams and drawings that I found on 4Chan, Above Top Secret, Dr. Tarrin Lupo, and Harmony Mandela Effect's videos. I believe that if countless individuals, particularly medical professionals, agree that the body has changed, then we must take a closer look and not write these things off as false memories. I will continue to share with you some of the public comments that I received on my video blog at the end of this chapter. I think you may be surprised at the experiences that the Mandela-affected have. We will now continue with the next set of anatomy changes, the internal organ structure of the human body.

Internal Organs-

Another Mandela Effect where people are noticing a change is the internal organs themselves. A change in the size and location of our internal body structure and other anomalies have risen to the surface on a frequent basis since the Mandela Effect has come to my attention. One peculiarity incorporates the exact size and area of the liver; folks claim that it has doubled in size and appears misshapen. This change of the liver has also drawn attention to the kidney. Mandela Effect sufferers have noticed that it now seems to be pushed up behind our ribs, due to the larger size of its neighboring organ.

Also, the size and volume of the stomach are seemingly too small to countless individuals, leading to a lack of appetite, weight loss, and heartburn. Another debate is that people remember the esophagus lying vertically before, yet now it curves to meet the stomach. Information about the volume and length of the intestinal tract present in the human body is also baffling and dividing scholars and doctors abroad due to this phenomenon. The layout of the intestines also seems quite sloppy, and taking up way too much real estate in our mid-sections. Even the appendix appears to be in a new location peculiarly hanging from the large intestine. I have even had multiple people recall their blood type being different from the one that they have today.

How does one's blood type change without some genetic modification or manipulation?

Other health concerns have arisen, including some who say that the size and shape of their lungs have become smaller and more compact in size, leading to shortness of breath and other strange breathing issues. Some of these matters have been complaints of chest pain, and symptoms that mimic cardiac arrest, such as numbness and suffering down one or both arms. Neck pain, headaches, and general

The Mandela Effect

lethargy are running rampant for those who are experiencing the anatomy changes physically. Now that we have skimmed the surface of the Mandela Effect on human organ changes, I will share with you some current diagrams of human anatomy and digestive systems for your review:

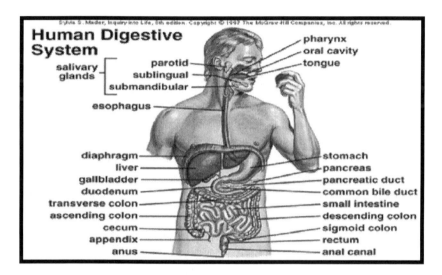

Human Digestive System Drawing
Author- Reddit

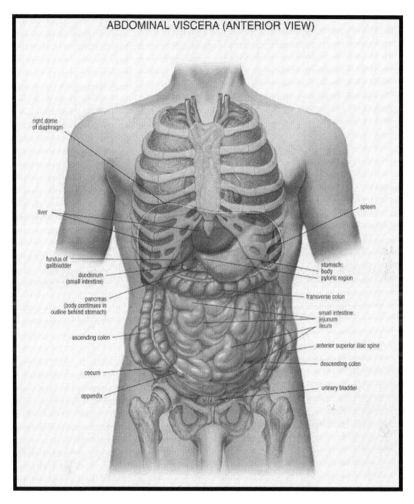

**Abdominal Viscera (Anterior View)
Retrieved from- Cryton Chronicles**

Diving further into artistic renditions, which cannot commonly be manipulated as easily as digital art can, I discovered an early drawing of the human anatomy, found dating back to Greek Culture. Look closely at the artistry in this next image and reflect on how the internal structure is expressed here. What do you see?

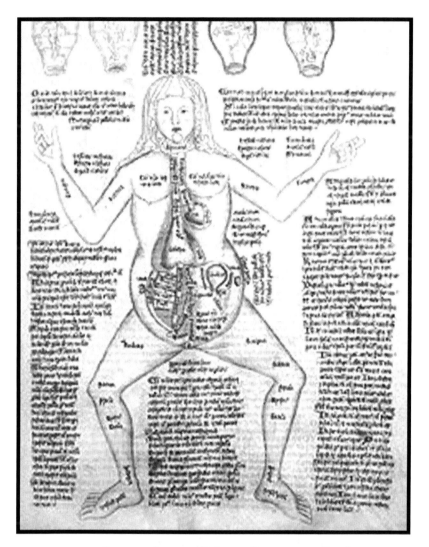

**Image of an early rendition of anatomy findings
Author- Wellcome Library, London**

Ancient Greek knowledge of anatomy and physiology underwent great changes and advances throughout early medieval times. Over time, Greek medical practices expanded through a continually developing understanding of the functions of organs *and structures in the body. Phenomenal anatomical observations of the human body were made, which have contributed towards the understanding of the brain, eye, liver, reproductive organs, and the nervous system. – **Wikipedia.org***

I took the same approach that I had before by searching YouTube for additional videos based on the Mandela Effect anatomy changes to see if anyone else had covered the changes linked to the internal organs. I was surprised to stumble upon yet another video presentation from Byron Preston, aka Harmony Mandela Effect.

Mandela Effect Human Anatomy Has Changed!: https://youtu.be/IK95OJI-pLI

Harmony begins to point out the changes that he noticed inside of the human body. The first observation that he makes is regarding the shape, size, and location of the stomach. Harmony mentions that a portion of the stomach is now contained within the rib cage, and it did not used to be located there before. There is also a remarkable change in the volume of the small intestine; it appears to take up the majority of where the stomach used to be. The size, shape, and angle of the esophagus also seem to have shifted. It looks as though the esophagus is now at a severe angle, making it difficult for food to digest. The size and volume of the human liver have also appeared to have doubled in size, if not more. It was evident that Harmony was shocked by these changes, and he openly asked his audience to share their memories of the human internal organ system. With permission from Harmony, I would like to now share with you some of the public comments that he received on his YouTube video presentation.

Justice Forall-

Thank you. My family thought I was crazy when I was freaking out looking at the intestines on a diagram, a couple of weeks back. I distinctly remember them being zig-zagged, and now they are a jumbled mess. I also think our brain stem moved as well as our spine; it looks different than what I remember seeing.

Bob Bant-

Ribs, Skull, Spine, Brain, Stomach, Liver all wrong. That temporal bridge is such an obvious change. It is all good though no big deal. I believe it is critical to be nice to all now. Has anyone noticed people who do not experience the effect are somewhat hostile? This place seems more hostile. To me, it means we are here to spread good will and help spread positivity.

Jack Skellington-

Since when is the liver LARGER than the stomach? I remember men having one less rib. I always thought it was a pretty big clue.

Anton Nym-

OMG. This is a mind-blower. Of course, I remember the original stomach configuration. I hadn't realized it has changed! Unbelievable. I had to check Wikipedia and other sources. The stomach was never partially behind the ribcage before. It was right in the middle and somewhat shaped like a kidney bean. This wedge-shaped stomach, I don't even recognize! The thing is, I was at my Gastroenterologists last month, and there was a poster on the wall with the "normal" configuration. I didn't even think twice about it. So, it seems to me that this change has happened within the last 30 days.

GypsyKing-

Yeah, it's totally off by a lot -but not only the anatomy has changed on those illustrations, but also my mood has gone down a lot in the past year or so. I was more upbeat till some time ago than I am these days -and I don't even watch Tel-Evil-Sion, read the newspapers nor listen to the public radio broadcasting. No disturbing effects from the outside media at all, and I'm not in a good mood lately. It's not right; it feels like I was in another world these days. They are changing everything from the inside out.

kazz 13-

I have noticed getting certain pains from certain body parts (like the stomach) to change. Now if I'm running and I get a stitch, it's more on the side, and I have no idea why. Sort of like the body has changed.

Vondale Reynolds-

Definitely, the Mandela Effect: I am no Doctor, but I am a Radiology Technologist (24 years), and I can truly say this is not right. As for the stomach, it's not supposed to be located behind the rib cage, and I do recall it always being on the left, never in the center of the torso, many people do mistake the small intestine for the stomach. And where the heck is the bronchial tube? Why are all the ribs connected? ...No floaters?? Why are the lungs so small? So many questions, this is all wrong. But it is so hard to wake people up to this event.

Lanting Farming-

The fact that the heart is suddenly in the middle makes me definitely feel like I was going crazy. I can't imagine that my heart was always in the middle, because as a young kid, from the age of 7, I was sent to a doctor because of some noise against my heart, and he put the stethoscope on my left side of the breast. When I had a run or something, I felt my heart on the LEFT side. I know 100% sure that my heart belongs to the left side. And not the middle. Just as the size of the liver, it makes no sense being that big. I want to go back to our dimension where I was. It seems to me that we are slowly turning into totally other beings with maybe six fingers and toes, and large eyes, just like they describe aliens. Sometimes, I wish I'd never heard of this Mandela Effect because it can make you insane. Everything changes, and it is even more frustrating when nobody else seems to sense that everything is different as it was. I know for sure that CERN is behind all of this with the LHC-Toy....what more can we expect? Thank you so much for this video, I really appreciate it. It confirms what I thought, and it is a damn tough time for us being awake. Blessings from the Netherlands.

Rhea Blazer-

We know that our heart was on the left side because that is the reason why we put our wedding rings on our left ring finger, because of the vein that leads to our heart. We aren't going crazy here.

Mark Linenberger-

Definitely, the stomach is on the wrong side, and the liver was set lower.

Luke Gramith-

Our bodies and anatomy must've changed overnight because about two years ago, I noticed this bump on the back of my head. You can feel it for sure. I can't lay on my back on a hard floor without being in pain. It turns out it's connected to my spine. It definitely wasn't there before! But it's been around for a couple of years now. Also, my neck cracks constantly.

Donnie Brown-

I went to the E.R. about a year ago, thinking I had chest pain, but they determined that I just had an upset stomach. I remember arguing with the nurse when she told me that the stomach was way up in my chest, not lower, as I had always believed. Well, I sure felt foolish, but not anymore after seeing this video. Thanks for sharing.

TMCicuurd12b42-

Sounds like me. I was freaking out. I thought I had water in my left lung because it was making gurgling noises and there was definite movement resonating through my left chest. As my father had died of lung cancer, I was worried. I went to see my family doctor, and he laughed because it was simply my stomach making sounds because I was hungry. I was in disbelief. Doc, I've been hungry before, and I had noise coming from my belly, not my left lung. He then described that contrary to popular belief, the stomach is not entirely in your belly. I was like WTF? I've been hungry every day in my life for the better part of 40 years. Why all of a sudden would it shift from my belly to my left lung!!?

ToKnowIsToDie-

For me it's different. I'm no doctor, but the liver is too large, the stomach is too small and too high. The kidneys were lower. Lungs are not long enough; the ribs are off. The intestines were packed in like a coil, not all spaghetti like. Yeah, it's crazy off.

mattplus09-

The intestines look like someone didn't feel like taking the time and just shoved everything in. I hate to think of my food taking that mountain road!

Raphael Aegis-

I have old anatomy books that couldn't have been digitally changed as all information on the internet can be. I swear to God that our anatomy has changed. The alimentary canal was the man on the video says; it was straighter. Also, the stomach was larger and situated roughly where he says. Also, the liver is absolutely gigantic compared to what I remember; I know it was. The heart was situated more to the left and smaller. I don't even know what to say, and 90% of the planet seems oblivious. I am certain about what I have said.

Adam Colbert-

Want a perfect example (and example of "residue") of where the heart was in our previous timeline??? Look up "Pulp Fiction adrenaline shot scene." Also, it's so weird now... I remember being able to FEEL my heart thumping... You could put your finger in that little "hollow" part of the ribs, just inward and upward of the left nipple. Now, NOTHING!

Ron Willis-

I have studied anatomy, and it's different now to me as well. The stomach up on its end looks weird?

Human Heart:

Another conundrum that has crossed my path on this Mandela Effect journey is that thousands of individuals believe that the location, shape, and size of the human heart have changed. Let's look at what the current information on the heart is before we get into the changes:

The heart is a muscular organ in humans and other animals that pump blood through the blood vessels of the circulatory system. Blood provides the body with oxygen and nutrients, as well as assisting in the removal of metabolic wastes. The heart is located in the middle compartment of the chest.

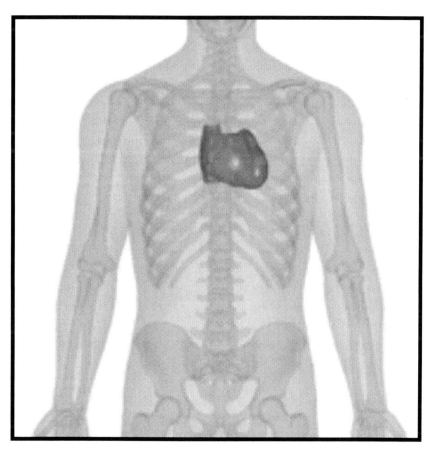

**The human heart is in the middle of the thorax, with its apex pointing to the left.
Author- BodyParts3D/Anatomography**

The human heart is situated in the middle mediastinum, at the level of thoracic vertebrae T5-T8. A double-membraned sac called the pericardium *surrounds the heart and attaches to the mediastinum. -**Wikipedia.org***

Since prehistoric times, people have had a sense of the heart's vital importance. Cave paintings from 20,000 years ago depict a stylized heart inside the outline of hunted animals such as bison and elephants. The ancient Greeks believed the heart was the seat of intelligence. Others believed the heart to be the source of the soul or emotions—an idea that persists in popular culture and various verbal expressions, such as *heartbreak*, to the present day.

- Microsoft Encarta

Now I want to talk a bit about the changes that I remember. The one thing that keeps repeating in my mind about the location of the heart reminds me of my days as a schoolgirl. Each morning before class, we had to place our right hands over our hearts to recite the pledge of allegiance. If our heart had always been in the center of our chest, then why on Earth would our teachers ask us to place our right hands over our hearts for our entire elementary school years? We had to start class this way, every single morning. How could I possibly forget this? I even recall a scene from the movie "Wayne's World" where Garth creates the shape of a heart with his hands directly on the left side of his chest. There are countless other examples of residual evidence such as this, but this is just to scratch the surface and reawaken your memories. So many other people recall the heart being on the left side.

What do you remember?

The Mandela Effect

**Barack Obama apparently does *not* have the Mandela Effect.
Author- Snopes.com**

Did you know that your heart is psychically located nearer to the focal point of your mid-section? Many individuals recollect that it had previously been found entirely under the left rib cage. Not to mention all the tattoos people have received over their hearts! Should the artists have placed their tattoos in the center of their chest instead? Some people also recall that the human heart itself used to be the size of one clenched fist—not two as it stands today. It appears our hearts have grown larger with the Mandela Effect. I find this fascinating and symbolic all at the same time.

The heart has always been a mystery—it drives us, it guides us, it saddens us, it keeps us alive...but has it physically changed?

You decide.

The Mandela Effect- Changes to the Human Anatomy:
https://youtu.be/w_h3YAgKESc

Now that we have reviewed the top Mandela Effect Anatomy changes, I would now like to share with you some of the comments I have received on the anatomical changes from viewers of my YouTube Channel with their permission. I want to thank everyone who has sent me feedback, personal stories, and residual evidence. I am always receiving new emails from people daily and will continue to update my research and tell your stories. I honestly could not have completed this project without your help, support and encouragement. Now let's get to the comments!

Joan Clayton-

Yep, it has to be connected. last night and the day/night before, I think a significant shift happened again because I got the headaches nightmares and disturbed sleep- waking up several strange hours of the evening feeling like I was in a "sick" body...my Hubby is feeling ill as well, just wondering if we are all going through the shifts/adjustments at the same time or if some of us M.E. er's are transitioning at different times? The first few waves of my personal transitioning/ascension I felt really the same ill/off feeling as now, but I wasn't consciously aware that it was a Mandela related thing at the time...I started paying attention to my physical symptoms instead of just blowing them off as a "flu thing" about a month ago. xo

Rita-

After ruling out other things/health problems, I am with the camp who thinks this is happening to our bodies because we have been moved into new bodies in a new timeline, and a new earth with different geography. Different astronomy/planetary sizes and positions have shifted, and our anatomy is different. We're experiencing symptoms because our reality change was so quick and we're not fully adjusted to it yet. Just look at the size of our liver—now twice as big as the old one, and our heart lungs and stomach and even intestines are in different places than we're used to, not to mention our skeletons have changed. Look at how different our rib cages are now! And our skulls and brains are different. Hang on. I think it's going to get worse before it gets better! Drink lots of bottled water and stay away from fluoride toothpaste!

Barbi Button-

Yes, Rita - stay away from all fluoridated water (most bottled waters are now too).

Mahima-

Human anatomy has changed also. The stomach is not where it used to be. My husband was having some cough issues, so I was looking at some medicine sites. And the human anatomy pic just popped up, and I was shocked to see that it was certainly not what it used to be it. I feel like I'm in another world. Thank you, Stasha, for talking about this and thank you for writing a book about this...I remember my teacher in school explaining things about organs some 15 years ago, and I distinctly remember where the stomach used to be...this is serious stuff. I told my husband, and he remembers nothing. He says I'm crazy to think it was somewhere else. Check out the human anatomy, guys. It's bizarre.

Integrative Options-

Yep, you are from where I remember too. Thank you for responding. And I have all the same anatomy memories too.

Kismet-

I've only just become aware of The Mandela Effect... Waking up every night about 2 am, feeling very, very sick! Liver hurting, hot flashes, and depression like you wouldn't believe. Awful. I just don't know what is going on...

Greg-

Mandela Effect: fact not theory, if I were only 99% certain, I would NOT say anything! I was aware of the Mandela Effect for two months before my ribs spine and skull changed on June 13-2016. I had a floating rib on my front left lower rib cage, never had four short ribs only attached to my spine, and the human spine was smooth, the human skull did not have a bone bridge under the temple!!! My friend had a human skull, and I would stare at it every day for several years, and in high school, I sat next to a skeleton in class, and I would play with it every day. If you do not know what a floating rib on your front lower rib cage is, then you are originally from here. A floating rib here is when one of the short back ribs are broken or detached from the spine, and it is painful—a floating rib where I came from is a short rib on the front lower rib cage attached with cartilage to the rib above it and is NOT painful. Every man has one on their front left lower rib cage, and every woman has one on both. There was an Old Testament story that explained why men had one less rib than women (God made Adam sleep and took one of Adam's ribs and formed Eve from Adam's rib).

Allan-

My body is different here; the heart is on the very left-hand side of the chest. In the other world, there, the liver is located on the lower left-hand side of the body, down to the waist. There are also ribs down there.

Charles-

Personal Mandela Effect is at one time I had a weak heart valve, and now my heart valves are fine. I now have an enlarged heart, and my heart valves have always been okay. It would take longer to explain, but that is it in a nutshell.

Derek-

My heart used to be right over on the left side of my chest, but everything changed one year three months ago, and now my heart is in the center of my chest. I have done a web-search and found a corporation website for a medical device. On this page of the site, I have left a link to a projection of the heart which is shown on the left side of the chest. I think this could be residual evidence:

http://www.denasusa.com/index/zones_of_device_application/0-109

Daryl-

Two years ago, I found a birthmark on my right cheek, yet my whole life I never had any blemish or anything on my face! CERN is maybe mixing not just one other dimension with ours but maybe many?

Nate-

I guess the heart is in the middle now... mine is still on the left. I remember doing CPR class on the left side too.

I want to thank every individual who sent me personal stories, comments and their own Mandela Effect stories. I will continue to share the stories and research I have received into written works so that no stone is left unturned. I am by no means an expert in the field, but with due diligence, I will continue to present the facts as I see them changing. The mysteries of the anatomical changes of the human body are becoming too vast to count at this stage. I have hundreds of pages of residual evidence from my YouTube subscribers, friends, and clients. We have only hit the tip of the proverbial iceberg on anatomy effects. Call this an introduction to Mandela Effect anatomy changes, as I will continue to update this project into a continuous series of books so that no individual human's story is left untold.

5. Astronomical Changes:

Just when we thought the Mandela Effect could not get any more peculiar, your mind will be blown in this next section about our physical universe. The first thing brought to my attention was in the field of astrology. Since I have a presence on the Internet for being a spiritual guide, I have a large number of clients who perform or follow astrology. I have never fully understood astrology, nor had the capacity to learn it, but I do trust the experts in that field.

I was appalled when I witnessed one of my client's monthly astrology videos on YouTube for the sign of Capricorn (The Astrological Symbol of the Western Zodiac). In her monthly description video for Capricorn, she presented an image of a creature that I had never witnessed with my human eyes before. The typical mountain goat that I always remembered seeing depicted by this astrology sign now had the tail of a mermaid! This new chimera of a beast was known as a Sea Goat, and apparently, it has always been this image in our reality. I have heard of various mythical creatures, yet have never in my life heard or even *seen* an artist's rendition of a SEA GOAT. Have you? I asked my client about her video, and she stated to me "Capricorn can be represented in both ways." I was not convinced, so I had to dive further down into the rabbit hole for research. What was this mythological creature? The Mandela Effect debate lies in whether Capricorn is represented by a Sea Goat or a Mountain Goat. Look at the Capricorn symbol as it is today on Wikipedia:

**The mysterious "Sea Goat" known as Capricorn.
Author- Wikipedia**

Still not convinced by Wikipedia and the modern image they presented, I dug a bit deeper into art history and found this next gem in the Vatican Library that blew my mind!

Unknown Artist, Aquarius, and Capricorn, Rome Book illustration in Miscellanea Astronomica, Bibliotheca Apostolica, XV secolo, Vatican City, Vatican Library.

The Vatican represents a goat with sea legs, yet, thousands of Mandela Effected individuals and I remember Capricorn being represented by a mountain goat. The Library of Congress, in the Thomas Jefferson Building, displays a mural reflecting Capricorn as a mountain goat, and I was even able to find this next image that dates to the 15th century. Thankfully, this artist remembers it the way that I do:

Zodiac sign of Capricorn in a 15th-century manuscript, by e-codices.

*So where on Earth did the beloved legs of the mountain goat that we all remember go? I believe it is imperative to define what exactly a sea goat is meant to be. Online, I linked it to a constellation figure known as Capricornus. "Capricornus /ˌkæprɪˈkɔːrnəs/ is one of the constellations of the zodiac. Its name is Latin for "horned goat" or "goat horn" or "having horns like goats," and is commonly represented in the form of a sea-goat: a mythical creature that is half goat, half fish." -***Wikipedia**

As per usual, Wikipedia also presented conflicting results on yet another link related to Capricorn on their site under the subheading of Astrology. To me, this is a form of residual proof that something has changed when you read the information closely. "**Capricorn** (English pronunciation: /kæp.rɪ.kɔːn/) is the tenth astrological sign in the Zodiac, originating from the constellation of **Capricornus.** It spans the 270–300th degree of the zodiac, corresponding to celestial longitude. Under the tropical zodiac, the sun transits this area from December 22 to January 19 each year, and under the sidereal zodiac, the sun transits the constellation of Capricorn from approximately January 16 to February 16. In astrology, Capricorn is considered an earth sign, negative sign, and one of the four cardinal signs. Capricorn is said to be ruled by the planet Saturn. Its symbol was based on the Sumerians' primordial god of wisdom and waters, Enki with the head and upper body of a mountain goat, and the lower body and tail of a fish. Later known as Ea in Akkadian and Babylonian mythology, Enki was the god of intelligence (gestú, literally "ear"), creation, crafts, magic, water, seawater and lake water.

The mountain goat part of the symbol depicts ambition, resolute, intelligence, curiosity, but also steadiness, and the ability to thrive in inhospitable environments while the fish represents passion, spirituality, intuition, and connection with the soul. Individuals born between December 21 to January 19 may be called Capricornian. Capricorn is third and last of the earth signs in the zodiac, the other two being Taurus and Virgo."

I also noticed that when I researched the Virgo symbol in astrology, that it had also changed! Virgo used to be represented by the Harvest. She was a mother figure who lovingly held the cornucopia, full of the abundant harvest. Now, when you research Virgo, she is a water bearer, yes... a water bearer! And her infamous cornucopia is also missing. There are other depictions of Virgo that are found on Wikipedia, depicting her now as a young maiden with a stalk of wheat, but still no cornucopia.

> *"In classical antiquity, the cornucopia (from Latin cornu copiae) or horn of plenty was a symbol of abundance and nourishment, commonly a large horn-shaped container overflowing with produce, flowers or nuts.* **(Wikipedia, 2017)**.

There is infinite symbolism in these words that I found through these astrological links. If you are one of the people experiencing the Mandela Effect that believe witchcraft or black magic are connected, then here is a ton of logical proof for you. Let me break down a few things that immediately came to my attention. First, it was the mention of the different zodiacs, tropical or sidereal. We will touch more on this topic of the various constellations in the next section, as NASA has already changed the dates of our astrology signs, as well.

The next thing that caught my attention was the use of the word "negative" about describing the sign. I never recall any sign being known as negative before, as all humans have different sides, right? Then I was floored when I realized the connection to Saturn and Capricorn, as I had always known it to be ruled by Pluto. The dots began connecting when I conducted more research on Saturn and what that specific planet represented. Many cultures attach Saturn to the being known as Satan. So, perhaps this is where the "negative" sign has a relation.

Next, they continued to mention the links to the Sumerians and Enki, and I knew something "fishy" (pun intended) was going on here. This is how they twisted the sea goat into the mix. They connected Capricorn to Enki, the fish god. Enki is not as innocent as they have attempted to present here on Wikipedia. He is also often compared to "the beast out of the sea" from the book of Revelation in the Holy Bible (Yet another connection to Satan/Saturn). If there is any chance that witchcraft has been at play, then a Babylonian god would most certainly be a clue. Not to mention a god who represents "magic."

Just when I thought astrology could not get any more confusing, at the beginning of my research on the Mandela phenomenon, after 2,000 years, NASA announced that they had apparently changed our astrology signs. You will not believe the peculiarity of this article:

YOUR ASTROLOGICAL SIGN HAS SHIFTED: NASA UPDATED THE ZODIAC SIGNS FOR THE FIRST TIME IN 2,000 YEARS:

September 2nd, 2016:

The stars are a database of information, from overcoming struggles to forming healthy relationships, to improving your character, and so much more. This knowledge has been used since ancient times in various cultures, and now it has been passed to you.

NASA Updates Star Signs for the First Time in 2,000 Years:

The stars have shifted due to the wobbling effect of the Earth caused by the sun and the moon over the last two to three thousand years. The sky is different than what the ancient astrologers viewed. Now, the constellations are off by about a month.

Therefore 86% of people are living under the wrong horoscope.

Presenting the All New Horoscopes:

- Capricorn: Jan. 20-Feb. 16.
- Aquarius: Feb. 16-March 11.
- Pisces: March 11-April 18.
- Aries: April 18-May 13.
- Taurus: May 13-June 21.
- Gemini: June 21-July 20.
- Cancer: July 20-Aug. 10.
- Leo: Aug. 10-Sept. 16.
- Virgo: Sept. 16-Oct. 30.
- Libra: Oct. 30-Nov. 23.
- Scorpio: Nov. 23-29.
- Ophiuchus: Nov. 29-Dec. 17.
- Sagittarius: Dec. 17-Jan. 20.

Check that List Again-

This will blow your mind. There are thirteen zodiac signs.

Actually, that is not a new revelation. **Ancient Babylonian astrologers** discovered these thirteen constellations yet omitted Ophiuchus so that the signs will be divided equally into twelve due to the 360-degree path of the sun.

Well, this tells us one thing: government conspiracies are not a new phenomenon.

Ophiuchus: The 13th Sign-

The Ophiuchus constellation features a man grasping a snake and is more commonly known as **Serpentarius, the Serpent Holder**. This sign has been **adapted into many ancient cultures, but in essence, is the same being. (Satan/Saturn/Serpent)**. In Greek mythology, Ophiuchus is identified as being Asclepius the healer and **son of Apollo**. Asclepius held **power to resurrect the dead**. An enraged Hades persuaded Zeus to assassinate Asclepius with a lightning bolt. **In death, Asclepius was placed as a constellation in the stars.**

For the Sumerians, this figure was known as the god Enki. The Egyptians recognize this sign to be a man named Imhotep who lived in Egypt around the 27th century BCE. In the Bible, these attributes belonged to the Hebrew man Joseph, son of the forefather Jacob. Imhotep is the Egyptian name for Joseph.

Retrieved from:
http://thespiritscience.net/2016/09/02/your-astrological-sign-has-shifted-nasa-updated-the-zodiac-signs-for-the-first-time-in-2000-years/

The pattern here is becoming quite clear, is it not? As you can see, there just may be something to the concept and theory of spiritual warfare connected to the Mandela Effect. Babylonian magic is not a joke, nor is witchcraft. It is severe and can have tremendous consequences attached to it, particularly if the magician is not properly trained in their craft. Could the Mandela Effect have been caused by a secret society of Magicians? We shall see what unfolds as this mystery unveils itself.

The next peculiar astronomical change links to the entire Universe itself, and our location in the Milky Way Galaxy. Thousands of researchers have memories that we were previously located in what is known as the Sagittarius arm on the outer edge of the Milky Way Galaxy. Today, if you research our location in the Universe, we are now apparently located in what is known as the Orion Spur or Orion arm of the Milky Way galaxy. This is nearly on the opposite side of the galactic core from where we were before! I did the research myself and confirmed that this is our new planetary location. How could the Earth move so far without us noticing?

I checked Reddit to see what other researchers had found and stumbled upon a fascinating post regarding this planetary change:

Baked says:

"I saw a thread similar to this before but nothing with too much evidence to back some of the "misremembering." Ok, so I've been watching Neil Tyson videos on YouTube for quite a few years now, and I've watched all his episodes of Startalk. The guys awesome... and I have personally always thought we were in the Sagittarius Arm because I've heard Neil Tyson say it before. It took me a while to find where he says it in one of his videos... But I finally found it. On Startalk episode 4 at 37:55 (on Netflix) Neil Tyson says, "From Manhattan NYC, North America..." and continues all the way to the... "Sagittarius Arm." Now as "history" says we are in the "Orion Spur"... ALWAYS have been, either in the "Orion Arm" or the "Orion Spur." And the distance between the two (Sagittarius Arm and Orion Spur) is at least 1,000+ light years away from each other minimum. (I'm no astronomer, so this is my estimation by just given the fact that the Orion Arm is 10,000 light years long and by looking at the "Milky Way Galaxy Map") That's way too far just to be wrong... Now how does Neil Tyson "misremember" where we are in the Milky Way? He's one of the top Astrophysicists in the world. Of course, he could be... and I guess just IS apparently wrong. But that's surprising to me that he would not know where we are in the Milky Way. Does anyone else remember ever learning this same thing from another source? Or do you remember the Earth always being in the Orion Spur?

Here's a link to Startalk episode 4 (start at 37:45) https://youtu.be/zNSSY0n36Jw

Here is a possible Carl Sagan video also saying we are in the Sagittarius Arm. (I say possibly because I don't know what Carl's voice sounds like, but the video is labeled Carl Sagan. It almost sounds like a "robot- voice." I'm on my phone right now with no headphones, and my speaker is horrible... sorry in advance if I'm wrong)

Carl Sagan Link: https://dotsub.com/view/9063c8e8-e2f7-43b1-bae9-a43bbb204890

Retrieved from:
https://www.reddit.com/r/MandelaEffect/comments/4zwzk6/neil_degrasse_tyson_wrong_our_location_in_the/

NEIL DEGRASSE TYSON - REMEMBERS THE SAGITTARIUS ARM.
Retrieved from: National Geographic Channel

I certainly find it peculiar that an accomplished scientist and astrophysicist such as DeGrasse, would just misremember the location of our Universe, don't you? I decided to hop back over to YouTube to see if I could find any video presentations on this change in our galaxy position. I located yet another video display from Harmony Mandela Effect:

I FOUND the OLD EARTH!!! Mandela Effect:
https://youtu.be/fytKtTYLwZQ

Byron begins the video by asking his viewers if they remember our location in the Milky Way Galaxy, and how many light years away we are from specific planets in the universe. Byron then displays a famous artists representation of space, made by Jerry Lodriguss in 1997. The picture is marked with a yellow line stating, "You are Here":

You are Here, 1997
Author- Jerry Lodriguss

After displaying this picture, Byron recalls that this is where he remembers our planet is located on the outskirts of the Milky Way Galaxy. NASA and the majority of the astronomy world, claim that we are located in an area of the galaxy known as the Orion Spur. Now the distance from the sun to the center of the galaxy is 25,000 light years, and the distance to the outer stretches of the universe is also 25,000 light years away. In other words, our solar system is now smack dab in the middle of the Milky Way Galaxy! Yet Byron remembers us being on the Sagittarius arm, way on the outskirts of the Milky Way Galaxy, not in the center!

It seems as though we have completely switched positions in the galaxy to the polar opposite of where we were before. Now you cannot see the Earth from the center of the Milky Way Galaxy. We have an observation shadow of the galactic core. Planets located on opposite sides of the core should be invisible to one another. The galactic center should be almost impossible to travel due to a considerable black hole and neutron stars. Now things have shifted remarkably. The only explanation that Byron could obtain for this change, was a theory known as singularity. In physics, singularity is defined as a point at which a function takes an infinite value, especially in space-time when the matter is infinitely dense, such as at the center of a black hole. Similar to string theory, if a planet is attached to another via an invisible line, then they could also change locations in unison. I understand that the theory that we have possibly switched locations in the Milky Way Galaxy seems unbelievable and unimaginable, so I have decided to share with you the public comments that Byron received on his video so that you can see he is not alone in his theory.

Public Comments on Galaxy Changes:

Jacob Hollis-

I'm 21, and I still remember that our solar system was on the outskirts of the milky way.

Jay Hollis-

I believe CERN is doing something. I truly believe it. Call me crazy but...

Oliver Johnson-

I'm 13, and I'm pretty sure we were told we on the outskirts recently.

Salt & Pepe-

Well if you think of solar systems as mysterious, mystical bodies then this could be abnormal phenomena in every solar system over time meaning we are entering a new Age of Enlightenment. We are overdue a catastrophe and its most likely not going to happen; we are entering Aquarius.

Anson Meyer-

Does this mean.... we're on Earth-2 basically? Or are we dealing with a timeline alteration, because I have been clearly noticing changes around us?

AVA Magnetic Levitation-

Do more like this you are triggering memories in others. And you may have already played a role in a much larger action yet to come. I worked the night shift for eight years around the time of change the night sky has never looked the same from back then to now. It's almost a sign in the face of the affected to see things so much changed when other say always been like this or must have always been this way without even thinking too hard about it.

It still all could be an AI in quantum time control fixes twin planets and a shuffle to keep us all confused or save itself from us. Maybe it did start changing things to take control but we had a backup time mark or way to stall it back in time before it comes but by its quantum laws it would have and know all time and paths before we even made it in the first place like a time paradox once it's out of the box no way to ever close it off from us and may be no way to stop it reaching farther in space and time than us and even be able to send its self much information and self-understanding from future evolutions of its self and all at once, Like Skynet but self-aware in divisions of time so small we can't even measure or conceive with more knowledge and understanding than we could ever understand with the size and current evolutions of our brains, You do seem to be very right about the kind of people most affected or noticing it. 'The self-inflicted human suffering here just for $ makes me sick in ways I cannot say for now.

Jeremy Kin-

Revelation 21:1 "a new heaven and a new earth."

DarkestStar Gamez-

I have noticed changes not only about the earth but myself like I can predict things in my dreams, and if something bad is coming up then it starts to look like everything is shaking, I'm a completely different person from who I was and like, I have seen people with Illuminati like tattoos following me.

SlothyyJr-

No, I know we were on the outskirts. I remember NASA Officials saying we can see the Andromeda galaxy because we are on the outskirts. Same with teachers telling me that.

Draftyoffical-

I don't think we moved to a mirrored earth. I think we moved to a mirrored universe.

TrueXV-

It would not be possible that we were formed if there was indeed an alternate universe in which we were in the Sagittarius arm. It is important to accept the truth, one scarier than what you suggest, as humans our memories, even collective, are extremely unreliable. Welcome to the phenomenon of confabulation.

TheCgOrion-

Could this be the reason why our sun seems to be whiter than it used to be? Produces an intense burning feeling on your skin, even when it isn't higher temperature.

Christopher Wade Kirkland-

We are heading to the center.

FredyEX-

Hmm well, I remember the outskirts hmm...

Magical Mia-

Unless the center is a black hole type of thing and that sucking is what is creating that swirling arm effect... We USED to be on the outside arm, but it could be that like water going down a drain the "particles" in the swirly arms are circling the hole and being drawn into the center slowly. Same galaxy. Same planet, Just closer to the center.... oh. I've noticed the stars and the sun and moon being different too. Everything is moving...

Versace Diaper-

I remember being on the outskirts, but also near the middle, do you think people could've merged to become a new person, the way some new animals have.

Abir Roy Chowdhury-

It's not called singularity!! It's called quantum entanglement.

Geometry dash Darkwing-

I think we have switched again... My whole family and I woke up today feeling so strange like something was different, then our upstairs neighbors came downstairs to us, and they felt the same feeling, I asked my friends, they said the exact same thing. It's really strange...

Lesley Medina-

So, something crazy happened to me yesterday, I watched this video yesterday & so I wanted to see and try out something, I wanted to see if our plants are merged or something. So, I myself remember it being on the arm of Sagittarius, I remember this clearly because I was taught about this in school since a young age, so then I searched on Google where is the solar system located in the Milky Way and it showed me that I'm in the Orion, I was confused!!!!! And so, I asked my boyfriend what he remembers about our solar system being located at, and he said the same thing as I did, so I told him to search it up & so he did, this is what amazed me... for him, it showed up that he was still in the Sagittarius arm. So, I thought to myself how in the world is he in my dimension as I'm in his at the same time?!! I need someone to help me grasp this!

Edward Sheets-

I must be from another Earth or time myself. I remember so many things different. I honestly didn't realize how much till I started researching quantum theory and string theory, at first, I found an overwhelming amount of information on subjects like the double slit experiment and quantum entanglement. Almost as soon as I started my deeper investigation, it seemed like almost overnight so much of the information websites even YouTube videos jus vanished. Like I was getting too close to the truth and was placed in an environment where the information and explanations just disappeared. We seem just to be scratching the surface of information I found in abundance before. Like there was a data reset. Any ideas or thoughts are really welcome.

Myfairytalelife-

I don't usually post comments on YouTube, but this video intrigued me. I've always been fascinated by astrology ever since a very young age [I'm 27 now] so hearing some of the things mentioned in the video has been slightly eye-opening. I've always remembered learning that Earth was located on the Sagittarius arm at the very edge of the milky-way galaxy. Hearing that now we are located on the Orion's arm at the center of the milky-way feels strange. I don't know what to make of this information. I feel as if I got fired from a job I never even worked at.

Titania-

I typed in our solar system in galaxy, and a pic of the solar system was at the end of the arm -_

Octavio Carles-

Sorry to correct the effect I think is called Quantum entanglement is a physical phenomenon that occurs when pairs or groups of particles are generated or interact in ways such that the quantum state of each particle cannot be described independently of the others, even when the particles are separated by a large distance.

Andaluza-

I thought that I was very knowledge of our solar system since I was a child, my father taught me about it, and we use to watch the sky together at night. I work the graveyard shift, and I look at the sky every night. Some time ago before I looked into the Mandela effect. I realized it that the sky and the stars looked different, they appeared brighter, closer and I could see more stars. I thought that my eyesight had improved because I changed my diet to healthy eating. I was shocked to learn that I wasn't the only one. This is awesome!!!!

Multidimensional-Jump-

I do remember there being a black hole in the center of this galaxy. But yeah, I do remember us being at the end of the universe not near it's center. I do remember there being only JFK and Lincoln being assassinated. I do remember doing a rune reading about my soul group to see if I had 1 and I had four people 1 being my ex's name. Then later on someone did an OBE near me and then I did another rune reading and got her name the OBE woman. Now the reading I did today says there is one other woman making six instead of 4, then five souls. Me and three women than four women, and now five women. The other living Mandela effect woman now is somewhere here in NW Ohio, and I'm here Toledo, Ohio. I want to meet her since I caused her existence hard to explain but I do know for a fact I gave advice to something to change something and didn't believe it would do anything seeing how I woke up and I was still here and our Universe was still here. Now you are telling me it's different. Weird I don't care who thinks I'm nuts but I know what I did. It was for this or rather the old earth. So, it doesn't suffer anymore. But didn't it worked, I thought nothing happened plus thought I daydreamed it up more or less. But now huh. All these days ago when I knew something big was going to happen and these Mandela effects videos are making me deeply wonder WTF is going on. This so not a joke this comment I am 100% serious. Weirrrrrd.

TSXYBEAST-

I was big into astronomy from 95 to 2001, and we were located in the Sagittarius arm, I am 100 percent positive!!!

Corbro101 DGAF-

I also remember being out far on the stretch of the arm. Pretty much the first pic you showed. You're not crazy!!

It is apparent that Byron is not alone in his theory that we have shifted positions in the universe. While the public comments seem to be varied in response, there is clearly a level of deception surrounding Earth's galactic location. What memories do you have of the position of our planet in the Milky Way Galaxy?

What Happened to our Sun?

Another change that has come to the surface since the Mandela Effect was born, is the visible changes to our Sun itself. The sun no longer seems to shine the golden hues that it once did in the past. Now when we look outside, we are met with a HUGE blazing bright white sun, that feels too hot and almost artificial in nature. Many people theorize that if we have indeed changed location in the universe, then why wouldn't our sun change too?

Individuals have contacted me to say that the position of constellations has changed from the view that they had always accessed them. Others say that the place of the moon has made an 180-degree flip from where it once laid on their skylines, and some have even claimed that they have seen it move and rotate, which should be impossible. We were taught as children that the moon was in a fixed position, that it never moved, so now why the change? We now know that NASA has updated our astrological signs, which are linked to constellations, so could this be another connection?

Speaking of NASA, I researched the topic of an artificial moon, when I noticed the changes in the sun. I recorded a YouTube video and wrote a quick blurb about my research on an interesting patent that NASA had created, I think you will find this connection quite intriguing:

NASA has put a patent on an artificial sun- Mandela Effect: https://youtu.be/qe8jvOPmTpA

> *I am currently in the process of editing a chapter regarding the Mandela effects that have arisen in our atmosphere recently. The sun seems to appear artificial in nature. It is almost TOO BRIGHT, like an over-sized flashlight of some sort. After my recent research about NASA's patent on an artificial sun satellite, I knew the sun being reflective in nature, was no longer just a conspiracy myth.*
>
> *These atmospheric changes go beyond our Sun and have stretched out to our Moon, Stars,*

Constellations, and the Clouds... Oh man, don't even get me started on the clouds. Between the chemtrail created manufactured cross-shaped clouds, to the blanket-like pillow clouds, rippling like waves through our curiously orange-pink skies. I have been taking a collection of photos over the past two months of the skies that surround my homes in both Norway and California. Please enjoy the variations of the colors, structures and even patterns in some of the formations. The Sun is certainly looking like it has undergone some transformations. The exact origin to me is unknown. But someday soon, we will have answers.

Here is the link to the NASA patent:

https://matrixbob.wordpress.com/2016/04/11/nasa-patents-from-the-1950%E2%80%B2s-1960%E2%80%B2s-anti-gravity-artificial-sun-levitation-and-simulating-the-sun/ow

Now I will share with you some of the content that I found disturbing in this patent:

Simulating the sun, Patent holders NASA, Bausch & Lomb and more, (Mar 20, 2016).

Patents from 1950's to 1960's, NASA and other contributors are deep into advanced technologies well beyond what they show us, their rocket charade and all their programs are a front, there is much more going on than we can imagine, some of these patents are decades old, anti-gravity tech has been around since the late 1800's but well hidden from us, delve deep into the patents on artificial sun, levitation, anti-gravity and anything you can think of and look below at the other related patents, mind blowing inventions you never could imagine and some several decades old, if some of these work we many people could be fooled by simulations/simulators, antigrav orb tech, etc. Means for simulating certain environmental conditions of outer space:

https://www.google.com/patents/US3010220 Space chambers
https://www.google.com/patents/US3064364 Space simulator
https://www.google.com/patents/US3187583
<u>Levitation Patents</u>

NASA Patents from the 1950's to 60's & from other contributors are deep into advanced technologies well beyond what they show us, and there is much more going on than we can imagine. Anti-gravity tech has been around since the late 1800's and delves deep into the patents on the artificial sun, levitation, anti-gravity & anything you can imagine. Here are other related patents & mind-blowing inventions that are several decades old. Means for simulating certain environmental conditions of outer space Simulating the sun, Patent holders NASA, Bausch & Lomb & more Space chambers, Space Simulator: https://www.google.com/patents/US3187583

Levitation Patents:

http://ntrs.nasa.gov/search.jsp?R=19710020816

https://www.google.com/patents/US3010220

https://www.google.com/patents/US3064364

http://www.wildheretic.com/electric-sun-mechanics/

http://www.google.com/patents/US33252...

https://www.google.com/patents/US3239...

Retrieved from: http://beforeitsnews.com/space/2016/04/nibiru-was-patented-by-nasa-in-the-1950s-2497794.html?utm_source=dlvr.it&utm_medium=twitter&utm_campaign=beforeitsnews

The comments section from my YouTube video were as intriguing to read through as the NASA patents themselves:

art H-

The sky and the sun have been looking weird or a few months now, for me anyway. I have the ME as well, so I don't know if this is part of it, because a lot of people I know don't notice anything.

spoiled me-

WOW! Felt connected like remembering taking pictures and the emotion, have seen in other videos around the world, is a Universal Fractal (the Schiavoni Fractile), after

the quantum jump to the parallel Universe is all over. Welcome to the Multiverse. You have the eye of Horus showing for you, go and read The Kybalion. I salute you.

unimatrix 001-

I remember something about that. I have a memory about NASA putting a mirror in space. What they are trying to do is making days longer. So Before Sunset mirror reflects the sunlight back to the ground and day would be longer for 2 hours or something like that. However, NASA said this was a failure. They were trying to put a giant mirror in space.

Pretty Planet-

Patterns are an exact description of our clouds. I've noticed these as well, for possibly the past four years how odd that I too started taking pictures of clouds like jigsaw puzzles, large perfect circles as you've shown in another video, and what I call clone clouds looking like each other. Oh, how about the shadow clouds that attach themselves to other clouds. Well yes, there seems to be definite changes in a lot of our sky.

Wendimac-

I have started to really notice these things myself just recently, pertaining mainly to brand names and movie lines being 'different,' and the clouds being 'different'… About the sun, my son, who was autistic, and hypersensitive to changes in his environment, started staring at the sun during the spring of 2015, until he passed away over the Easter weekend. At the time, I thought it was another 'behavior,' possibly related to the cataract that he had in one eye. But now that I think about it, he was staring at it as if looking for something...and he was almost obsessed with it...and I have to wonder if he had noticed something different.

NoblenessDee-

HI, Stasha, I have a lot of research done on the world map using google earth ... feel free to use anything in any way. BUT GET THIS ... Scarabperformance has just blown the doors off everything. He has proven all kinds of the TV/Ads Mandela Effects using the US patent office website ... it is ground breaking. it is a power shift ... it is a must see ... Everybody needs to help push this out there, he has already been hacked, they are not happy .... :) But I am. Hugs to all.

Roudy One-

Good to see you back don't let the trolls scare you off girl :)

Salted Seeder-

Doesn't NASA have Tesla's sun simulator patent?

Stasha Eriksen-

I thought so. Poor Tesla, he must be rolling around in his grave!

It is apparent that I am not alone in my observations of our changing Sun.

What exactly have you been up to NASA? Speaking of NASA, since they seem to be the center of attention throughout this chapter of the book. I want to touch upon another Mandela Effect that has many researchers including myself utterly baffled. The first topic of debate is what many refer to as "the Moon landing," but the problem is, there was apparently much more than just one manned mission.

Oh yes, that is correct, there have been six manned missions to the Moon recorded in history as of 2017. Yet thousands remember there only being one mission, with "one small step for man, and one giant leap for mankind" nothing more. Loads of conspiracy theories have popped up involving Stanley Kubrick and a code he apparently left in his movies (most notably The Shining) telling the truth about what NASA had forced him into creating, which was the infamous moon landing footage itself. But let us stick to the Mandela Effect in this book, we must get into the six manned missions, involving 12 men who stepped foot on the moon.

A moon landing is an arrival of a spacecraft on the surface of the moon, including both manned and unmanned (robotic) missions. The first human-made object to reach the surface of the Moon was the Soviet Union's Luna 2 mission, on 13 September 1959. The United States has consistently been in a "space race" with other nations ever since this day. The United States' Apollo 11 was the first manned mission to land on the Moon, on July 20th, 1969. There have been six manned U.S. landings (between 1969 and 1972) and numerous unmanned landings, with no soft landings happening from 22 August 1976 until 14 December 2013. To date, the United States is the only country to have successfully conducted manned missions to the Moon, with the last departing the lunar surface in December 1972.

After the unsuccessful attempt by the Luna 1 to land on the moon in 1959, the Soviet Union performed the first hard (unpowered) moon landing later that same year with the Luna 2 spacecraft, a feat the U.S. duplicated in 1962 with Ranger 4. Since then, twelve Soviet and U.S. spacecraft have used braking rockets to make soft landings and perform scientific operations on the lunar surface, between 1966 and 1976. In 1966, the USSR accomplished the first soft landings and took the first pictures from the lunar surface during the Luna 9 and Luna 13 missions. The U.S. followed with five unmanned Surveyor soft-landings. The Soviet Union achieved the first unmanned lunar soil sample return with the Luna 16 probe on 24 September 1970, followed by Luna 20 and Luna 24 in 1972 and 1976, respectively. Following the failure at launch in 1969 of the first Lunokhod, Luna E-8 No.201, the Luna 17 and Luna 21 were successful unmanned lunar rover missions in 1970 and 1973.

Many missions were failures at launch. Also, several unmanned landing missions achieved the Lunar surface but were unsuccessful, including Luna 15, Luna 18, and Luna 23 all crashed on landing; and the U.S. Surveyor 4 lost all radio contact only moments before its landing. More recently, other nations have crashed spacecraft on the surface of the Moon at speeds of around 8,000 kilometers per hour (5,000 mph), often at precise, planned locations. These have been end-of-life lunar orbiters that, because of system degradations, could no longer overcome perturbations from lunar mass concentrations ("mascons") to maintain their orbit. Japan's lunar orbiter Hiten impacted the Moon's surface on 10 April 1993. The European Space Agency performed a controlled crash impact with their orbiter SMART-1 on 3 September 2006.

Indian Space Research Organization (ISRO) performed a controlled crash impact with its Moon Impact Probe (MIP) on 14 November 2008. The MIP was an ejected probe from the Indian Chandrayaan-1 lunar orbiter and performed remote sensing experiments during its descent to the lunar surface. The Chinese lunar orbiter Chang'e 1 executed a controlled crash onto the surface of the Moon on 1 March 2009. The rover mission Chang'e three was launched on 1 December 2013 and soft-landed on 14 December.

The U.S. robotic Surveyor program was part of an effort to locate a safe site on the Moon for a human landing and test under lunar conditions the radar and landing systems required to make a validly controlled touchdown. Five of Surveyor's seven missions made successful unmanned Moon landings. Surveyor 3 was visited two years after its Moon landing by the crew of Apollo 12. They removed parts of it for examination back on Earth to determine the effects of long-term exposure to the lunar environment. Now that we have covered what began this initial "space race" of unmanned moon landings let us now dig into the documented "manned" landings that NASA claim to have achieved in this next section.

A total of twelve men have landed on the Moon. Accomplished with two US pilot-astronauts flying a Lunar Module on each of six NASA missions across a 41-month period starting on 20 July 1969 UTC, with Neil Armstrong and

Buzz Aldrin on Apollo 11, and ending on 14 December 1972 UTC with Gene Cernan and Jack Schmitt on Apollo 17. Cernan was the last to step off the lunar surface. (Cernan sounds like CERN, hmmm). All Apollo lunar missions had a third crew member who remained on board the Command Module. The last three missions had a rover for increased mobility.

Plans for manned Moon exploration originated during the Eisenhower administration. In a series of mid-1950s articles in Collier's magazine, Wernher Von Braun had popularized the idea of a manned expedition to the Moon to establish a lunar base. A manned Moon landing posed several daunting technical challenges to the US and USSR. Besides guidance and weight management, atmospheric re-entry without ablative overheating was a major hurdle. After the Soviet Union's launch of Sputnik, Von Braun promoted a plan for the United States Army to establish a military lunar outpost by 1965. I have since discovered that von Braun was a former NAZI, but that is another story for another day. NASA kind of sounds like NAZI too, doesn't it?

After the early Soviet successes, especially Yuri Gagarin's flight, US President John F. Kennedy looked for a US project that would capture the public imagination. He asked Vice President Lyndon Johnson to make recommendations on a scientific endeavor that would prove US world leadership. The proposals included non-space options such as massive irrigation projects to benefit the Third World. The Soviets, at the time, had more powerful rockets than the United States, which gave them an advantage in some kinds of a space mission.

Advances in US nuclear weapon technology had led to smaller, lighter warheads, and consequently, rockets with smaller payload capacities. By comparison, Soviet nuclear weapons were much heavier, and the powerful R-7 rocket was developed to carry them. More modest potential missions such as flying around the Moon without landing or establishing a space lab in orbit (Kennedy proposed both to Von Braun) were determined to offer too much advantage to the Soviets since the US would have to develop a massive rocket to match the Soviets. A Moon landing, however, would capture world imagination while functioning as propaganda.

Johnson had championed the US manned space program ever since the Sputnik scare, sponsoring the legislation which created NASA when he was in the Senate in 1958. When Kennedy asked him in 1961 to research the best-manned space achievement to counter the Soviets' lead, Johnson responded that the US had an even chance of beating the USSR to a manned lunar landing, but not for anything less. Kennedy seized on Apollo as the ideal focus for efforts in space. He ensured continuing funding, shielding space spending from the 1963 tax cut, but diverting money from other NASA scientific projects. This last dismayed NASA's leader, James E. Webb, who perceived the need for NASA's support from the scientific community.

The Moon landing required the development of the large Saturn V launch vehicle, which achieved a perfect record of zero catastrophic failures or launches vehicle-caused mission failures, in thirteen starts. For the program to succeed, its proponents would have to defeat criticism from politicians on the left, who wanted more money spent on social programs, and on those on the right, who favored a more military project. By emphasizing the scientific payoff and playing on fears of Soviet space dominance, Kennedy and Johnson managed to swing public opinion: by 1965, 58 percent of Americans favored Apollo, up from 33 percent two years earlier. After Johnson became President in 1963, his continuing defense of the program allowed it to succeed in 1969, as Kennedy planned.

In total, twenty-four U.S. astronauts have traveled to the Moon. Three have made the trip twice, and twelve have walked on its surface. Apollo 8 was a lunar-orbit-only mission; Apollo 10 included undocking and Descent Orbit Insertion (DOI), followed by LM staging to CSM redocking, while Apollo 13, originally scheduled as a landing, ended up as a lunar fly-by, using free return trajectory; thus, none of these missions made landings. Apollo 7 and Apollo 9 were Earth-orbit-only missions. Apart from the inherent dangers of manned Moon expeditions as seen with Apollo 13, one reason for their cessation according to astronaut Alan Bean is the cost it imposes in government subsidies

Retrieved from:
https://en.wikipedia.org/wiki/Moon_landing

Yes, you heard that's right folks. Apparently, we have been to the Moon and successfully walked upon it more than six times. We have also had a total of twelve men that have supposedly walked on the Moon. Feeling inspired, I recorded my own video on the topic of our astrology changing to see what other people remembered and the response that I received may shock you!

Astronaut Buzz Aldrin, the lunar module pilot of the first lunar landing mission, poses for a photograph beside the deployed United States flag during an Apollo 11 Extravehicular Activity (EVA) on the lunar surface. Author- Apollo 11 Image Library

Now that we have reviewed the top Mandela Effect Astronomical changes, I would now like to share with you some of the comments I have received on these changes from viewers of my YouTube Channel with their permission. I want to thank everyone who has sent me feedback, personal stories, and residual evidence. I receive new emails from people daily and will continue to update my research and tell your stories. I honestly could not have completed this project without your help, support and encouragement. Now let's get to the comments!

The Bible and Astrology have changed?

https://youtu.be/uo-V3Sx3Dsk

markscott77777777-

Have you ever seen the movie Apollo 13? Tom Hanks. The real incident is my first memory.

Flat Earth Freedom w/ Lisa Elizabeth-

I remember Capricorn being a mountain goat going up a hill slow and steady.

Phil C-

Interesting video, thanks, I don't claim to have a deep knowledge of the Bible, and as you say it's been edited and censored to turn Christianity into a state controlled religion. I recall the Pesher documents discovered in the dead sea scrolls and the Nag Hammadi texts indicates that the original books were coded on four levels and the format reflected that coding, you probably know more about that than I do. When these things were altered under Constantine, and later, the coding no longer matches, so it's possible to see which texts are changed. I found the astrological changes by accident when I looked at my own sign Taurus, and it now is ridden through rough waves, Capricorn, as you say is now a water sign, the present age Pisces is also a water sign, as will be Aquarius. If as some say the earth has moved, then it follows that the zodiac will have changed in this reality, that perhaps may explain these zodiac change? Just guessing, it beats me!

Dragon Star Alchemy-

Hmm, interesting theory but I'm not a "Christian" anymore, but I noticed the effects...9-11 was always an inside job. It was "on earth." Trespasses (trespasses even makes more logical sense than debtors). Lion and the lamb...Capricorn was always just a goat. Yes, you're on to something Saturn is also Chronos (god of time) ... Personally, I think the Mandela effect is a ploy of the corrupt systems attempts to gaslight the "awake"/"conspiracy" people to distract from something else. While something is going on, I think most people out there are genuinely being deceived, joining in mass hysteria, or trying to jump in the hype and maybe deceive people. (I'm not accusing you of deceiving people.) Use your disCERNment to discover the truth.

alan brunt-

Sea-goat = Barack Obama in English Gematria. All Mandela effects are clues. Great video, thank you, and God bless you.

Girlface OneTwoFree Leslie-

Whaaaaaaaat? I'm a Capricorn, and I have NEVER seen a sea goat.... It's ALWAYS been the mountain goat... Always... I'm 35, and I as well have never known these things to be as they are now. So much is different, and I'm not afraid I am simply curious... I've only just found you here, so I'm going on a watchathon of your videos now! Also going to try to get some sort of connection/reading! Anyway, just super stoked to have stumbled across you here I feel good about it!

Paul-

I can't take the CERN/DWave theories seriously. There are, most likely, a lot of advanced civilizations in the Milky Way Galaxy alone (over 100 billion stars) that likely have technologies far more advanced than quantum computing. If quantum computers were capable of "breaking the universe" it would have happened by now. Just my own opinion.

Kaun Australis-

Edgar Mitchell is the Astronaut who believes in Aliens BTW. Also, Edgar Mitchell was one of the astronauts that heard the music from the dark side of the moon. I imagine they got quarantined just in case any pathogens or bacteria from the moon got into their bloodstream.

OldDood-

Stasha...Check out a website named: www.thelivingmoon.com

A friend of mine has built this up over the years. His name is Zorgon. John Lear is involved too. (Son of the creator of the Lear Jet). There is so much data there it will take days and days to try to go through all of it...

Alt Portal-

In my reality... of course, there were more than one. Were they real? Who knows. But, yes. I watched the landings and splashdowns of several. Do not recall the exact amount. I recognize the astronaut names. They quarantined them upon return... yes. I always found that to be creepy. Maybe since we were so pro-space shuttle for so long, those who were not around during the Apollo missions, just didn't know about them all or were even taught about them beyond the very first landing. I don't know.

Seth Roback-

Capricorn has always been a Sea Goat in my effect - it comes from the Babylonian deity of Enki who brought civilization to humans also it represents the rain season and flooding that happened in the Tigris and Euphrates and in the Nile where the Zodiac was developed.

David Sullivan-

I'm a Capricorn, and I only know it like a goat. Sea goat is new to me. So many things have changed, it's blowing my mind. my anchor is Jesus Christ. Us Christians need to pray for discernment and wisdom.

Wolfman Joe-

I am a Capricorn. When I first learned about what represented my sign years and years ago, I remember it was a "sea goat," because I remember thinking...."what is a sea goat"? However, I do remember Curious George having a tail, and other things, that now seem different.

Liv Watson-

I remember Capricorn as being represented both ways. As either a goat with a fish tail or just as a goat.

Nobody Baldwin-

Thank you for making this video. I did my birth chart today after not doing it for a few years. I am now a Pisces moon which I used to be a Capricorn moon, and my Venus is in Scorpio which used to be in Virgo. However, I believe this new chart to be more accurate than my original chart. Because it explains my psychic abilities. But the Mandela effect obviously is the reason my chart has changed.

Girlface OneTwoFree Leslie-

Also, IDK if you believe in this double sun anomaly but I always blew it off until I woke up one day and saw it for myself (I know it sounds beyond crazy to tons of people) so I took photos of it... Anyway, I feel like it sorts of fits in with the Mandela effect because it was just there one day... I'm sure I may get a negative response from someone, but I'm just throwing it in there!

Darmack Haller-

One time, this is true... I remember being taught this in school and reading it from the beginning days of the internet when you had to connect the phone to another phone. There have been dozens of mainstream documentaries saying one time, but yet suddenly seven?

And a little unrelated, but you had another video about anatomy changes. Not sure but I think I found another one. Back in 2013, they discovered the sixth layer to our eyes. I find this very hard to believe that it was there the whole time and no one noticed it. Doctors, scientists, and researchers tend to be obsessive compulsive; someone would have noticed it at least a decade, ago right?

Chante Moody-

I only ever heard about one manned landing on the moon.

Joe Hodgen-

Check out this video: https://www.youtube.com/watch?v=-gBQTjPxaK4

It is about the hollow moon ringing like a bell for an hour or something like that. Have a great day.

thomasvids1007-

The (NASA) images don't stick out. They look like they have the depth to them. The Netflix logo on my computer looks like it is on the surface of the screen. Has anyone noticed this?

David MacArthur-

Check this link out: http://www.nasa.gov/facts/Space/space_facts_archives.html

Rita Vapes-

Hubby says we landed two times on the moon. I only remember one. Xo ~Rita

Joe k.i Smithhiukhik.u have-

Real quantum effects are hard to find large amounts of reality residue.

Also, the reality residue from "real" quantum effects tends to disappear once it is pointed out on the internet. But keep looking into all these strange things. I think there is a connection, it all has to do with psychological conditioning.

Phil C-

I was 17 in 1969; I thought the Apollo missions the greatest thing we have ever done, I watched every mission, 11 to 17, and I was watching everything I could sometimes see at 2 or 3 in the morning. I knew all the astronaut's names and followed them through their training and missions on the moon, except for 13 of course. Due to lack of funding and loss of interest in the US. I was furious, I wanted to see us explore more, space stations, moon stations, ready to jump further, but no it was all squashed. It was only later I found out about the Van Allen belts and operation starfish exploding nuclear weapons in orbit made it 1000 times more toxic, the stories were that the Apollo missions went through the belts quickly, but try standing in a nuclear reactor for more than a few seconds and see what happens. Stanly Kubrick was alleged to have filmed the missions in a secret studio, and all the films and photos are fakes. There are some good documentaries on this, Kubrick tried to blow the hoax in The Shining and Steven King was furious that Kubrick had changed the book so much.

Kubrick's family say it was all hoaxed, we never went to the moon. Modern missions do not go above 350 miles, any higher and radiation starts to poison the astronauts, NASA themselves admitted that the radiation belts were 'impenetrable' in 2014. So did we go to the moon at all, was it all a political hoax? Even the film 'Interstellar' says so!

unimatrix 001-

I remember NASA sending people to the moon but not that many times.

gmswealth-

Those of us who are aware of the Mandela Effect are blessed or fortunate. If CERN did it, it backfired. If Satan did it…it backfired. If Quantum computing did it, it backfired. The Mandela Effect is a new beginning for the millions of us that are impacted by it. We can now discard our past failures and begin to be the people we were meant to be. We now have the collective power to change this timeline reality for the better.

The Mandela Effect let's all of us know; without question, that there is a GOD or GODS, and HE or THEY have given all of us, with the Mandela Effect, the physical evidence to prove beyond any reasonable doubt, that we are in another timeline reality with many of the problems that existed in our old timelines. Therefore, it is incumbent on all of us from these dead timelines to realize we are connected to this GOD or GODS and not to any religion; past or present, mindset, idiosyncrasy, ideology or whatever else we were taught to believe or decided to believe. I strongly recommend all of us; with the Mandela Effect, go forward from this point and enjoy this new experience, rather than fret over it. All of us working together can make this timeline reality a better one to live in for all the people that are living here.

We can now abandon the corrupt political parties and institutions we find here, the educational systems, banking systems, mainstream media outlets, economic systems, social systems and whatever else that we find here that does not benefit this planet and its inhabitants. Our past timelines are dead along with all the mistakes we made in them, either by our negligence, apathy, miseducation or ignorance. What we allowed to happen in our old timelines does not have to happen here. Accepting the reality that all of us with the Mandela Effect has been given a second chance, in real time is crucial to all of us working together for the first time in history to make the timeline reality on this planet a better world for all its inhabitants. We now have the collective power to make changes in our lives and this world, that was impossible before.

We can now demand; as a collective group, governments in all countries, endorse freedom and not liberties controlled by a corrupt few. This includes equality for all, equal protection under the laws, responsible citizenship, government accountability, criminal prosecution for government instigated false flag events, assassinations, environmental protection, so this planet will survive, an end to secret societies, black ops and all the other ills that were rampant in the timelines we came from. Together, the people with the Mandela Effect can correct all the mistakes of the old timelines we left behind so that the death, destruction, false religions and the madness that we left in our old timelines do not continue in this one.

Grape Ape-

I came across this video:
https://youtu.be/ZULrvHfWoew

I'm not saying I believe in all of what they are presenting here. I too believe the moon landing was faked, but the point is, this person also believed NASA never returned to the moon.

Catherine Garriga-

Orion will be the first space shuttle that will be able to make it out of Low Earth Orbit. NASA and Obama have stated this. No other ship has ever made in through the (Van Allen radiation belts (cross section) A radiation belt is a layer of energetic charged particles that is held in place around a magnetized planet, such as the Earth, by the planet's magnetic field.) Making it impossible by NASA's own words to have gone to the Moon.

https://www.youtube.com/watch?v=NlXG0REiVzE

Anastasia G-

I am surprised as you are! I find about 6 Apollo missions earlier, but I am happy to see someone who is surprised as I am. I do not believe that Americans ever were on the moon, not even once. And now, with more pictures, I believe even more that all of it b.s. First (always surprise me) where are the stars? Then, some pictures are taken in front of the lunar rocket? It is so small! How did three people, plus the Lunar Rover, food, restrooms, some sleeping space fit in this little "spaceship"??? Besides, the restrooms on a spaceship are still a problem even now. Also, I remember the movie, "Apollo 18", I watch it a while ago, (when everyone believes in one moon landing and 17 Apollo different missing). It is an obvious Scientific Fiction movie. But, now it is in a different perspective.

snow hero-

As for "did they make this up overnight?" from my experience in government work... they have a vault of "contingency" planning ready to change the story again and again infinitely as the "mission" dictates. It doesn't mean we don't get at the truth though. I experience M.E., but I do not give credit to CERN. I have seen the movie Wizard of Oz after all.

Queen t i Tay-

We are part of the Mandela fact because this is my first time hearing this.

J. Richard-

Does anyone remember that mini-series about the Apollo mission's a while back? Most of what I know of the missions are from that. To me, I always thought it was like 4-5 times. That the public got bored after Apollo 12, and they weren't really paying attention till Apollo 13's "Houston...we have a problem."

Stasha Eriksen-

Great point, and now if you look up that quote, it says "Houston, we've HAD a problem" what is really going on here?

Crafty Chrystal-

Stasha! Hi, from yet another reality. I understand there are at least four represented by the people here today. In mine, we landed three times…. and only three times. Apollo 11, 13, and 15. And the third time was when they took their rover with them. Big controversy when "Was it only a paper moon?" came out. Because there were so many discrepancies in the details… like, the rover could not fit into the lander, there was no room for the 2 astronauts to turn around in the module, the hole between the module and the capsule was too small for them to fit through with their suits on, the plans for the lander and the rover were "lost", so no way to really check. How do you lose the plans to the first vehicle to ever drive on the moon?! And lots of other stuff that was not explained, like why blue sky showed through the capsules window while they were in transit from the Earth to the moon. Most of the Earth believe we never really went there… just used special effects as they did in the movie "Capricorn One." Also, check out: "A Funny Thing Happened on the Way to the Moon."

Indigo Raven-

There have been zero manned missions to the moon and never will be.

whitehorse rider-

Holy crappie scrappy what the f@#$?????? One time, we have been to the moon ONE TIME. Well on the "old earth." we went to the moon ONE TIME... ONE TIME, One time...

I need a drink...

Julie Bond-

(This is) Insane!

Mankabec-

Mind blown, just found this out today. Huge history and science fiction buff as well. I'm in shock. All I knew about was one landing.

Chante Moody-

Well, I honestly never heard about any manned Moon landing but the first one. I don't think they ever landed a human on the Moon ever though (I think they are lying). If we went back so many times, why isn't this discussed and celebrated more? That would be an accomplishment to be very proud of, yet many (if not most) of us have never heard of any manned Moon landing but the one.

MISTYWORLD-

Wtf? This is news to me. Golf on the moon yeah right lol They are really playing the number game on this one omg Thank you for sharing This is mental.

Chante Moody-

I knew about them playing golf on the moon (which never made sense to me since I thought the lack of gravity should make that impossible). But, I assumed they were saying the astronauts played gold on the first trip to the moon because I only knew there was supposed to be one time they visited the moon. I always wondered why they wouldn't return to the moon if they successfully landed men on it before. SMH. I always that it was unlikely that they could get to the moon; I had doubts even when I heard about it as a small child. I also wondered how we knew the Earth was a moving ball, but I was told the image of Earth from space was proof. I assumed I was too stupid to understand. But, about seven years ago, I started finding out that a lot of things I was told were facts all my life was actually lies. So, since then I have been looking at things with a more open mind, and I have woken up to the fact that most of the things we are told are likely to be lies because there is no proof they are true. I realize the governments of the world lie to citizens, and the media lies and spreads propaganda. I don't believe NASA went to the moon even once! I don't believe the photo of Earth from space is legitimate, so I don't think it is proof of that the Earth is a spinning ball; I think it's probably flat.

Salted Seeder-

Just speed up the film of them walking on the moon, and you'll see its faked. Many, many, other things have been exposed. We never walked on the moon and never will.

Chica Cósmica-

I've watched the Conspiracy movie like three times before; I'm watching it now because with this ME things have changed even in old footages like in the footage of the JFK assassination, and I can't believe how much everything gets all messed up with this ME! Just watch the film again, and you'll see.

Salted Seeder-

Well, I guess the first lie wasn't enough, just look at these Astro-nots today who are having to live with this lie, you can see it in their eyes. If you have Netflix check out the movie called "the last man on the moon" and if you have time check out another one called "conspiracy theory moon landing" you can see reference in there that there was only one landing and proof that we never landed on the moon plus there's a new movie in theaters out right now about this hoax. But this is how our false force-fed reality works, and it just becomes our reality but not anymore for those with the eyes to see, and the spiritual blindness layers are being removed. The ME is the exposing of this false force-fed reality we've all been under and its part of an awakening process taking place, and it's caused by our creator. Conditioned people will claim it's CERN, quantum computers, etc. and that its merging timelines, parallel universes, dual earths, time machines blah blah blah... Those that can see clearly know what the cause is and what it all means.

Tony Chong-

Have you taken a look at the moon on Google Earth recently? It looks like a death star from Star Wars. It's also no longer a sphere. It has many facets to it like a fine cut gem. Thanks for posting this video.

Catherine Garriga-

Orion is a predominant, first key player in really allowing us to move human presence out of low Earth orbit," said Bill Gerstenmaier, associate administrator for NASA's Human Exploration and Operations Directorate.

Stasha Eriksen-

Maybe they just switched our location? Closer to Orion? Like were on the same planet per-se but in a new location?

damanofmen1958-

Hi, Stasha. IMO NASA has never landed on the moon. If you just look up John Lear Jr. (his father is the creator of the Lear Jet). Both he and his father claim that because of the Van Allen radiation belts', it is impossible for any living being to survive the cosmic rays. Just some food for thought. He also states that the monies all went to black ops for the military and was just a cover to gain tax payers money.

tcuff1333-

Land on the moon? You can see through the daytime moon if the moon were solid the negative space inside the crescent would be black as a shadow should be. Instead, the negative space matches the sky.

Neil deGrasse Pearshaped-

True, Star Wars death star programming.

Insanity is Sanity-

I found out about the Flat Earth theory two years ago, and the first thing I started researching was the Moon Landings, not just once but many, many times over the past two years. I KNOW I remember only landing on the moon ONCE. This is insane. (I say we landed on the moon once, but it's actually zero. I don't believe the landing(s) was real.

Sommerfugl-

Same here. Only one time and I don't believe we ever been there at all...

R Day-

+*Stasha Eriksen* Hiya Stasha, this is for your projects, books, and things. Hi, sanity if you see this. I read a few comments and really am interested now, saw several recalls only one trip. I don't trust NASA either as I see with f.e. (flat earth) belief I am also. Well I hope others report on this too I've always known up to Apollo 17. the dune buggy they took, right? # 11 was Armstrong the first man. But Russia was first to land a rover that returned with rocks, what I recall. I saw all this live btw. Well how interesting., I will watch your comments list here about the six additional missions including # 13 one. Did you guys not think the movie Apollo 13 was based on fact? This is "unusual," But the entire Mandela Elf attack is. Thanks, I am not editing my post here hope you understand my hieroglyphic typing, Rall.

Insanity is Sanity-

+*Gareth Wales* Hey man. Last night I was on the secret hangout show, and we were supposed to discuss the ME since it's my favorite topic right now, but we didn't get time :(I do seem to be one of the only few Flat Earthers talking about this, which I do find really strange. I have no idea what's going on, in my last vid I mentioned I don't think it's a natural thing because the changes are not consistent with natural changes. For example, if it was natural, why didn't the "Berenstein Bears" change to the "KLJGFDNLKGJDNG Bears." You know what I mean? All the changes seem like they have been thought through.

Stasha Eriksen-

For the record, I do not believe that we have EVER been to the Moon. This is why I referenced Stanley Kubrick. I have known all about the fake landings for years. I have this in my book. Also, that's why I mentioned it. The reason I was so blown away by this overnight discovery, is that myself and multiple other people have made documentaries and videos about the topic of "why have we never been back to the moon after all these years" this is why the seven extra missions are concerning. Not to mention that all of the photos from the other six missions look like they were all taken on the same day! Crazy times.

Joe Williams-

Here, here! We haven't even got by the cancerous radiation that you need to get by before you get to the moon. NASA has even said that without really realizing they have. Every picture taken from space has been photoshopped even the supposedly pics of the earth are computer graphics. They take billions and throw out bs. Just a glorified mafia is fleecing us all.

Caessarion-

I remember watching the ML, and I said to my dad it looked fake. So, in the 70's I ask, well why don't we go back, and that's when I learned about the other trips that weren't televised. We also used to talk about weird stuff like Hey Mikey; he likes it, he hates/eats everything. My thought about the fake ML was to take a picture of the earth. We only have 2-3 official NASA pictures of the earth; each is concerning. You may think there are too many people involved in a few, but there's no way for anyone to know. Everyone just does their job; no one even talks about it. First, I was looking for Nibiru in 2010 but found weird colorful stars using only binoculars and seeing a couple of vids. Then I looked for the curve of the earth, and couldn't find it. Then I look at the sun, it looks like it's moving, not me, and why does it get smaller at the last minute before it sets? I look around, and nothings moving. The water is flat and still, and I watch vids with Nikon 900 zoom on ships, it stays flat. I look at mountains that all lead up, from one range to another, 100's of miles away, when does it curve?

intx123-

I did a high school report about the 1 and only moon landing that seemed fake. I remember their suits were too primitive and the radiation in space would kill them. They needed ridiculous amounts of lead in the suit. The shadow angles were off like there was more than one light source and the flag had waves, but there is no wind on the moon.

alex giraldo-

All those images are CGI. Nasa is a hoax!

(Reply) Stasha Eriksen-

NASA= Never A Straight Answer

Gahri Smee-

I just watched a documentary today that said Pluto was shown to have frozen nitrogen over ice and I watched the same documentary before that showed Pluto was rock asteroid dust and ice, I remember vividly seeing it, so I know something is up.

ma he-

Virgo is carrying water! This is crazy!

(Reply) Stasha Eriksen-

I know, now if you google Virgo, it now says she is a young maiden with a stalk of wheat! No longer a Mother holding the cornucopia of the harvest!?

In conclusion, I am certainly not alone in believing that NASA has faked one or all of the Moon Landings. Many people ponder on if they have manipulated our star signs for some hidden reason as well. However, science will go to great lengths to discredit anyone who opposes their "evidence." We may never actually know how deeply this deception goes, but together we can expose and present our collective memories to one another. If we can continue to support each other's beliefs and allow for other opinions and outside of the box thinking, then perhaps we will get answers sooner than we had initially thought.

How many times do YOU remember us going to the Moon?

Do you believe we have EVER been there?

How can we explain why so many people remember things differently?

6. Bible Changes:

Just when I thought the Mandela Effect could not get any more peculiar, I was presented with what I consider to be the most remarkable change caused by this effect. Researchers and Bible scholars have discovered that words in the Holy Bible have changed. Yes, that is right... the Bible has changed! I have been in a Theology degree program since 2011, to obtain a Master's degree in Divinity eventually. Over the course of my studies, I have run across many elements of the Holy Bible that have simply felt "off" to me, yet I could not put my finger on it.

In my degree program, we study various translations of the Bible, to have a deeper grasp on the meanings of each text. I have noticed that there are slight variations to these different translations, but the changes presented to me with the Mandela Effect have gone beyond a few word switches. The changes related to the Mandela Effect and the Holy Bible appear quite cryptic in nature as if there is a code involved of some sort. I do not want to frighten anyone with the research that I am going to present in this chapter, as I honestly consider the Mandela Effect to be a positive thing, but we have to bring these changes to the surface.

The first change presented to me by a client was at the beginning of the King James Bible. The change that not only myself but thousands of other Bible scholars have noticed is an insert at the start of the King James Bible. This supplement, commonly referred to as an insert, was a dedication to Queen Elizabeth the 1st, by King James. I always thought that the Bible was not to be added to, nor taken away from, this would be blasphemy. What authority allowed Elizabeth to be featured at the start of our Holy Bible? Not to mention the most commonly used version of the Bible that exists today. I encourage you to grab your Bible and follow along with me for the remainder of this chapter.

Let us now begin with the beginning...

Queen Elizabeth in the Bible? Insert added to King James Edition:

1769 King James Bible Introduction-

TO THE MOST HIGH AND MIGHTY PRINCE JAMES, [BY THE GRACE OF GOD,] KING OF GREAT BRITAIN, FRANCE, AND IRELAND, DEFENDER OF THE FAITH, &c.

The Translators of the Bible wish Grace, Mercy, and Peace, through JESUS CHRIST our Lord.

GREAT and manifold were the blessings, most dread Sovereign, which Almighty God, the Father of all mercies, bestowed upon us the people of [England], when first he sent Your Majesty's Royal Person to rule and reign over us. For whereas it was the expectation of many, who wished not well unto our [Sion], that upon the setting of that bright

[Occidental Star], Queen [Elizabeth] of most happy memory, some thick and palpable clouds of darkness would so have overshadowed this Land, that men should have been in doubt which way they were to walk; and that it should hardly be known, who was to direct the unsettled State; the appearance of Your Majesty, as of the [Sun] in his strength, instantly dispelled those supposed and surmised mists, and gave unto all that were well affected exceeding cause of comfort; especially when we beheld the Government established in Your Highness and Your hopeful Seed, by an undoubted Title, and this also accompanied with peace and tranquility at home and abroad.

But among all our joys, there was no one that more filled our hearts, than the blessed continuance of the preaching of God's sacred Word among us; which is that inestimable treasure, which excelleth all the riches of earth; because the fruit thereof extendeth itself, not only to the time spent in this transitory world, but directeth and disposeth men unto that eternal happiness which is above in heaven.

Then not to suffer this to fall to the ground, but rather to take it up, and to continue it in that state, wherein the famous Predecessor of Your Highness did leave it: nay, to go forward with the confidence and resolution of a Man in maintaining the truth of Christ, and propagating it far and near, is that which hath so bound and firmly knit the hearts of all Your Majesty's loyal and religious people unto You, *that Your very name is precious among them: their eye doth behold You with comfort, and they bless You in their hearts, as that sanctified Person, who, under God, is the immediate Author of their true happiness. And this their contentment doth not diminish or decay, but every day increaseth and taketh strength, when they observe, that the zeal of Your Majesty toward the house of God doth not slack or go backward, but is more and more kindled, manifesting itself abroad in the farthest parts of [Christendom], by writing in defence of the Truth, (which hath given such a blow unto that man of Sin, as will not be healed,) and every day at home, by religious and learned discourse, by frequenting the house of God, by hearing the Word preached, by cherishing the Teachers thereof, by caring for the Church, as a most tender and loving nursing Father.*

There are infinite arguments of this right Christian and religious affection in Your Majesty; but none is more forcible to declare it to others than the vehement and perpetuated desire of accomplishing and publishing of this work, which now with all humility we present unto Your Majesty. For when Your Highness had once out of deep judgment apprehended how convenient it was, that out of the Original Sacred Tongues, together with comparing of the labours, both in our own, and other foreign Languages, of many worthy men who went before us, there should be one more exact Translation of the holy Scriptures into the [English Tongue]; Your Majesty did never desist to urge and to excite those to whom it was commended, that the work might be hastened, and that the business might be expedited in so decent a manner, as a matter of such importance might justly require.

And now at last, by the mercy of God, and the continuance of our labours, it being brought unto such a conclusion, as that we have great hopes that the Church of [England] shall reap good fruit thereby; we hold it our duty to offer it to Your Majesty, not only as to our King and Sovereign, but as to the principal Mover and Author of the work: humbly craving of Your most Sacred Majesty, that since things of this quality have ever been subject to the censures of illmeaning and discontented persons, it may receive approbation and patronage from so learned and judicious a Prince as Your Highness is, whose allowance and acceptance of our labours shall more honour and encourage us, than all the calumniations and hard interpretations of other men shall dismay us. So that if, on the one side, we shall be traduced by Popish Persons at home or abroad, who therefore will malign us, because we are poor instruments to make God's holy Truth to be yet more and more known unto the people, whom they desire still to keep in ignorance and darkness; or if, on the other side, we shall be maligned by self-conceited Brethren, who run their own ways, and give liking unto nothing, but what is framed by themselves, and hammered on their anvil; we may rest secure, supported within by the truth and innocency of a good conscience, having walked the ways of simplicity and integrity, as before the Lord; and sustained without by the powerful protection of Your Majesty's grace and favour, which will ever give countenance to honest and Christian endeavors against bitter censures and uncharitable imputations.

The Lord of heaven and earth bless Your Majesty with many and happy days, that, as his heavenly hand hath enriched Your Highness with many singular and extraordinary graces, so You may be the wonder of the world in this latter age for happiness and true felicity, to the honour of that great GOD, and the good of his Church, through Jesus Christ our Lord and only Saviour.- END.

This insert is riddled with cryptic and esoteric messages. I have found various scholars who have broken down the symbolism of this insert, but what parts of it do YOU find odd? I especially thought the part about our "tender and loving nursing Father" to be the strangest of them all. Since when does a father nurse anyone? At what stage did the King James Bible become the American standard when it contains a cryptic supplement like this at the start?

I wanted to look even deeper into what the King James Bible represented, and at what stage the words lost their meaning? I was pleasantly surprised when I discovered a book titled King James Onlyism: A New Sect, By James D. Price. In chapter 6, on page 121 I uncovered an extensive list called "WORDS WHICH HAVE CHANGED IN MEANING" in the KJV. I was shocked to see how many word changes existed in this Holy Book. His research was published in hardcover book form on Amazon in 2006, so this is certainly not a new revelation. I will now share with you this detailed list of word changes for your review.

"The following list of over 500 archaic and obsolete words and phrases has been prepared in order to help the average reader understand more readily the meaning of the King James Version. Of course, not all of these words and phrases are inappropriate in all contexts, and hence each expression is followed by a list of those passages in which misunderstanding is likely to occur. In each instance, the first word or phrase represents the King James text. This is followed by a dash to separate it from the more meaningful alternative. Where there is more than one alternative, each is given, and the corresponding passages are listed.

Though this list of archaic and obsolete words is not exhaustive, it does, however, provide the reader with a handy reference to most expressions which are likely to produce difficulty in comprehending the meaning of the King James Version."

WORDS WHICH HAVE CHANGED IN MEANING:

- abased -humbled, Mt. 23:12; Lk. 14:11; 18:14

- abide -await, Acts 20:23

- abroad -outside, Deut. 24:11 -in the street. Jer. 6:11

- accursed -devoted, Josh. 6:17, 18 (3 times); 7:1 (2 times), 11, 12 (2 times), 13 (2 times), 15; 22:20; 1 Chr. 2:7

- accuse -slander, Prov. 30:10

- acquaintance -acquaintances, Lk. 2:44; 23:49; Acts 24:23
- admiration -wonder, Rev. 17:6
- advanced -appointed, 1 Sam. 12:6
- advertise -advise, Num. 24:14 -reveal to, Ruth 4:4
- affections -passions, Gal. 5:24
- again -back, Mt. 27:3; Lk. 14:6
- show again -show, Mt. 11:4
- against he come -to meet him, Ex. 7:15
- agone -ago, 1 Sam. 30:13
- allege -prove, Acts 17:3
- allow -approve, Lk. 11:48; Rom. 14:22; 1 Thes. 2:4 -accept, Acts 24:15 -know, Rom. 7:15
- almost -soon, Ps. 94:17
- amazement -terror, 1 Pet. 3:6
- ambassage -embassy, Lk. 14:32
- amend -mend, Jn. 4:52
- ancient(s) -elder(s), Ps. 119:100: Is. 3:2, 5, 14; 9:15; 24:23; 47:6; Jer. 19:1 (2 times); Ezek. 7:26; 8:11, 12; 27:9
- anon -immediately, Mt. 13:20 -straightway. Mk. 1:30

- answer -defense, 2 Tim. 4:16
- apothecary -perfumer, Ex. 30:25,35; 37:29; Eccl. 10:1
- approve -prove, 2 Cor. 6:4; 7:11
- armholes -elbows, Ezek. 13:18
- artillery -weapons, 1 Sam. 20:40
- assay -essay, Acts 9:26; 16:7; Heb. 11:29
- assay(ed) -attempt(ed), Deut. 4:34; 1 Sam. 17:39 -ventures, Job 4:2
- at -near to, Ex. 19:15; Num. 6:6; Ezek. 44:25
- attendance -attention, 1 Tim. 4:13
- audience -hearing, Gen. 23:10, 13, 16; 1 Sam. 25:24; Neh. 13:1
- badger's skin -sealskin, Ezek. 16:10
- bands -pangs, Ps. 73:4
- barbarous people -barbarians, Acts 28:2
- barren -bereaved, Song 4:2; 6:6
- base -lowly, 1 Cor. 1:28; 2 Cor. 10:1
- because -that, Mt. 20:31
- bed -couch, Song 1:16 -litter, Song 3:7

- behaved -stilled, Ps. 131:2

- behind -lacking, Col. 1:24

- belied -denied, Jer. 5:12

- belly -body. Ps. 31:9

- bethink themselves -lay it to heart, 1 Kgs. 8:47; 2 Chr. 6:37

- betimes -early, Gen. 26:31; 2 Chr. 36:15; Job 8:5; 24:5; Prov. 13:24

- bewitch -amaze, Acts 8:9, 11

- bewray -betray, Mt. 26:73

- bewrayeth -revealeth, Prov. 29:24

- bishopric -office, Acts 1:20

- bloody -blood-thirsty, Ps. 5:6; 26:9; 139:19

- botch -boil, Deut. 28:27,35

- bottle -jar, Ps. 119:83; Jer. 13:12 (2 times)

- bottles -wineskins. Mt. 9:17; Mk. 2:22; Lk. 5:37, 38

- bottles of wine -wine-skins, Josh. 9:13; 1 Sam. 1:24; 10:3; 16:20; 2 Sam. 16:1; Job 32: 19

- bottles of wine -heat of wine, Hos. 7:5

- bound -landmark, Hos. 5:10

- bowels -heart, Gen. 43:30; 1 Kgs. 3:26; Ps. 109:18; Song 5:4; Is. 16:11; 63:15; Jer. 31:20; Lam. 1:20:2:11: Phlm 12, 20; 1Jn. 3:17 -hearts, Col. 3:12; Phlm. 7 -affections, 2 Cor. 6:12; Phil. 2:1 -anguish, Jer. 4:19 -tender mercies, Phil. 1:8

- box -jar, Mt. 26:7; Mk. 14:3; Lk. 7:37

- branch -song, Is. 25:5

- breaking up -breaking in, Ex. 22:2

- breeding -possession, Zeph. 2:9

- brigandines -coats of mail, Jer. 46:4; 51:3

- brought up for -bore unto, 2 Sam. 21:8

- bruit -report, Jer. 10:22; Nah. 3:19

- bunches -humps, Is. 30:6

- by -beside, Num. 6:9 -by reason of Num. 6:11 -against, 1 Cor. 4:4

- by and by -immediately, Mt. 13:21 Mk 6:25; Lk. 17:7; 21:9

- by that -before, Ex. 22:26

- careful -anxious, Lk. 10:41; Phil. 4:6

- carefully -anxiously, Mic. 1:12 -diligently, Phil. 2:28

- carefulness -anxiety, 1 Cor. 7:32; 2 Cor. 7:11

- carriage(s) -baggage, Judg. 18:21; 1 Sam. 17:22; Is. 10:28; Acts 21:15 -the things that ye carried about, Is. 46:1

- certified -told, Esth. 2:22

- certify -advise, 2 Sam. 15:28

- chapiter(s) -capital(s). Ex. 36:38; 38:17, 19, 28: 1 Kgs. 7:16 (3 times), 18 (2 times), 19, 20 (2 times), 31, 41 (2 times), 42; 2 Kgs. 25:17 (3 times); 2 Chr. 3:15; 4:12 (2 times): Jer. 52:22 (3 times)

- chapmen -traders, 2 Chr. 9:14

- chapped -parched, Jer. 14:4

- charge -burden, 1 Tim. 5:16

- charger -platter, Mt. 14:8, 11; Mk. 6:25, 28

- charity -love, 1 Cor. 8:1; 13:1, 2, 3, 4, 8, 13; 14:1; 16:14; Col. 3:14; 1 Thes. 3:6; 2 Thes. 1:3; 1 Tim. 1:5; 2:15; 4:12; 2 Tim. 2:22; 3:10; Tit. 2:2; 1 Pet. 4:8; 5:14; 2 Pet. 1:7; 3 Jn. 6; Jude 12: Rev. 2:19

- charitably -in love, Rom. 14:15

- cheek -chastisement, Job 20:3

- chief -chiefs, Lk. 19:47; Acts 19:31; 25:2; 28:17

- chief men -corners, Is. 41:9

- Chronicles -chronicles, 1 Kgs. 14:19, 29; 15:7, 23, 31; 16:5, 14, 20, 28; 22:39, 45; 2 Kgs. 1:18; 8:23; 10:34; 12:19; 13:8, 12; 14:15, 18, 15:6, 11, 15, 21, 26, 31, 36; 16:19; 20:20; 21:17, 25; 23:28; 24:5; 1 Chr. 27:24

- clean -completely, Josh. 3:17; 4:1, 11; Ps. 77:8; Is. 24:19

- clouted -patched, Josh. 9:5

- old cast clouts -worn-out clothes, Jer. 38:11. 12

- cockatrice' -adder's, Is. 11:8; 14:29; 59:5; Jer. 8:17

- collops of, maketh -gathereth (fat on his loins), Job 15:27

- comely -seemly, Ps. 33:1; Eccl. 5:18; 1 Cor. 7:35; 11:13 -stately, Prov. 30:29

- comeliness -majesty, Ezek. 16:14

- communicate unto -share with, Gal. 6:6; Phil. 4:14,15; 1 Tim. 6:18; Heb. 13:16

- communication -companionship, 1 Cor. 15:33 -fellowship, Phlm. 6

- compass -circle, Prov. 8:27 -encompass, Jer. 31:22

- compasseth -searchest out, Ps. 139:3

- concluded -shut up, Rom. 11:32; Gal. 3:22

- confection -spice, Ex. 30:35

- conscience -consciousness, 1 Cor. 8:7; Heb. 10:2

- consult(ed) -counsel(ed), 1 Kgs. 12:6, 8; 1 Chr. 13:1; 2 Chr. 20:21; Heb. 5:7; Ps. 62:4; 83:3, 5; Ezek. 21:21; Dan. 6:7; Mic. 6:5; Hab. 2:10

- consumption -destruction, Is. 28:22

- contrary -different, Ezek. 16:34 (2 times)

- convenient -needful, Prov. 30:8 -fitting, Rom. 1:28; Eph. 5:4; Phlm. 8

- conversation -in the way, Ps. 37:14; 50:23 -citizenship, Phil. 3:20 -life, 1 Pet. 1:15 -way of life, 2 Cor. 1:12; Gal. 1:13; Eph. 2:3; 4:22; Phil. 1:27; 1 Tim. 4:12; Heb. 13:5,7; Jas. 3:13; 1 Pet. 1:18; 2:12; 3:1,2, 16; 2 Pet. 2:7; 3:11

- convert -turn, Is. 6:10

- corn -grain, (multiple references too many to include here) -grainfields. Mt. 12:1

- cornfloor -grainfloor, Hos. 9:1

- cornets -castanets, 2 Sam. 6:5

- corrupt, are -scoff, Ps. 73:8

- countervail -compensate for, Esth. 7:4

- countries -country. Lk. 21:21

- country, a -the country, Jn. 11:54

- cover with a covering -pour out a drink offering, Is. 30:1

- cracknels -cakes, 1 Kgs. 14:3

- crouch -prostrate. 1 Sam. 2:36

- cuckoo -sea gull, Deut. 14:15

- cunning -skillful, Gen. 25:27; Ex. 35:35; 36:8; 38:23; 1 Sam. 16:16, 18; 1 Kgs. 7:14; 1 Chr. 22:15; 25:7; 2 Chr. 2:7 (2 times). 13,14 (2 times); 26:15; Song 7:1; Is. 40:20; Jer. 9:17; 10:9

- curious -skillfully woven, Ex. 28:8, 27, 28; 29:5; 35:32; 39:5, 20, 21; Lev. 8:7

- curse -devoted thing, Josh. 6:18

- damnation -condemnation, Mt. 23:14; Mk. 12:40; Lk. 20:47; Jn. 5:29; Rom. 3:8; 13:2; 1 Tim. 5:12 -judgment. 1 Cor. 11:29

- damned -condemned, Mk. 16:16; Rom. 14:23; 2 Thes. 2:12

- darling -dear life, Ps. 22:20; 35:17

- darts -weapons, 2 Chr. 32:5

- daysman -umpire. Job 9:33

- deal -part. Ex. 29:40; Lev. 14:10, 21; 23:13, 17; 24:5; Num. 15:4, 6, 9; 28:9, 12 (2 times), 13, 20 (2 times), 21,

28 (2 times), 29; 29:3 (2 times), 4, 9 (2 times). 10, 14 (2 times), 15

- dearth -drought. Jer. 14:1

- deceived -enticed, Jer. 20:7 (2 times)

- decline -turn aside, Deut. 17:11

- decline after many -turn aside after a multitude, Ex. 23:2

- defense -gold, Job 22:25

- delicately -cheerfully, 1 Sam. 15:32 -luxuriously, Lk. 7:25

- delicates -delicacies, Jer. 51:34

- deliciously -wantonly, Rev. 18:7,9

- delight -delicate living, Prov. 19:10

- denounce -declare, Deut. 30:18

- desolate -condemned, Ps. 34:21,22 -found guilty, Is. 24:6

- destroy -hold guilty, Ps. 5:10

- devils -satyrs, 2 Chr. 11:15

- direct their work -give them their recompense, Is. 61:8

- disappoint -confront, Ps. 17:13

- discomfited, be -become tributary, Is. 31:8

- discover(ed)(eth)(ing) -disclose(d)(th), 1 Sam. 14:8,11 -strippeth, Ps. 29:9 -reveal, Prov. 18:2 -uncover(ed)(ing), Lev. 20:18; Deut. 22:30; 2 Sam. 22:16; Ps. 18:15; Is. 3:17; 57:8; Jer. 13:26; Ezek. 13:14; 16:57; 23:10,18 (2 times),29; Hos. 2:10; Nah. 3:5; Hab. 3:13 -removed, Is. 22:8

- ditch -reservoir, Is. 22:11

- divide -distribute. Neh. 9:22

- doctors -teachers, Lk.2:46

- doctrine -the message. Is. 28:9

- dote -be fools, Jer. 50:36

- doubtful -anxious, Lk. 12:29

- dragons -jackals, Job 30:29; Ps. 44:19; Is. 13:22; 34:13; 35:7; 43:20; Jer. 9:11; 10:22; 14:6; 49:33; 51:37; Mic. 1:8; Mal. 1:3 -sea-monsters, Ps. 148:7

- drams -darics, 1 Chr. 29:7; Neh. 7:71, 72

- draught -drain, Mt. 15:17; Mk. 7:19

- duke(s) -chief(s), Gen. 36:15f, 29f, 40f et al.; Ex. 15:15; 1 Chr. 1:51 et al., -princes. Josh. 13:21

- durable -stately. Is. 23:18

- ear(ed) (ing) -plow(ed) (ing), Gen. 45:6; Ex. 34:21; Deut. 21:4; 1 Sam. 8:12; Is. 30:24

- early -earnestly. Ps. 63:1; 78:34; Prov. 1:28; 8:17

- earnestly -carefully. Lk. 22:56 -steadfastly, Acts 23:1

- earring -ring, Gen. 24:22,30,47

- Easter -the Passover, Acts 12:4

- either -each. Lev. 10:1; 2 Chr. 18:9

- emerods -tumors, Deut. 28:27; 1 Sam. 5:6,9, 12; 6:4,5,11,17

- ensue -pursue, 1 Pet. 3:11

- entreat(ed) -treat(ed), Mt. 22:6; Lk. 18:32; 20:11; Acts 7:6, 19; 27:3; 1 Thes. 2:2

- was entreated of -granted his prayer, Gen.25:21

- environ -compass, Josh. 7:9

- erred through, have -reel with. Is. 28:7 (2 times)

- estate -council, Acts 22:5

- estates -men, Mk.6:21

- ever, or -before, Dan. 6:24; Acts 23:15

- evidently -openly, Gal. 3:1 -clearly, Acts 10:3

- evenings -deserts, Jer. 5:6

- evil -trouble, Jon. 1:7 -bad, Jer. 29:17
- evil entreated us -dealt ill with us, Deut. 26:6
- exactors -taskmasters, Is. 60:17
- failed for -failed in looking for, Lam. 4:17
- fame -report, Mt. 4:24; 9:26, 31; 14:1; Mk. 1:28; Lk. 4:14, 37; 5:15
- fast -close, Ruth 2:8, 21, 23
- feebleminded -fainthearted, 1 Thes. 5:14
- feller -hewer, Is. 14:8
- fetch(ed) a compass -make (made) a circuit, 2 Sam. 5:23; 2 Kgs. 3:9; Acts 28:13 -turn about, Num. 34:5
- fillet -line, Jer. 52:21
- fit -appointed, Lev. 16:21
- fitches -spelt, Ezek. 4:9
- flagons -cakes of raisins, Song 2:5
- flagon(s) of wine -cake(s) of raisins. 2 Sam. 6:19; 1 Chr. 16:3; Hos. 3:1
- flags -reeds. Ex. 2:3, 5
- flanks -loins. Job 15:27
- flood -river, Josh. 24:2, 3, 14, 15

- floor -the threshing floor. Hos. 9:2; Mic. 4:12

- flowers -impurity, Lev. 15:24, 33

- folk in -nations for, Jer. 51:58

- forward -earnest, 2 Cor. 8:17 -ready, 2 Cor. 8:10 -eager, Gal. 2:10

- forwardness -earnestness. 2 Cor. 8:8 -readiness, 2 Cor. 9:2

- frankly -freely. Lk. 7:42

- fray -frighten, Deut. 28:26; Jer. 7:33 -terrify. Zech. 1:21

- free -willing. Ps. 51:12

- freely -for nought, Num. 11:5

- fretted -raged against, Ezek. 16:43

- froward -wayward, Prov. 4:24 -crooked, Prov. 8:8; 17:20 -perverse. Deut. 32:20; 2 Sam. 22:27; Job 5:13; Ps. 18:26 (2 times); 101-4; Prov 2:12. 15; 3:32; 6:12; 8:13; 10:31, 11-20; 16:28, 30; 17:20; 22:5

- frowardness -perverseness, Prov. 2:14; 6:14; 10:32

- furnished -filled, Mt. 22:10

- furniture -saddle, Gen. 31:34

- furrows -transgressions, Hos. 10:10

- galleries -in the tresses, Song 7:5

- gender -breed, Lev. 19:19
- glistering -glistening. 1 Chr. 29:2; Lk. 9:29
- go beyond -transgress, 1 Thes. 4:6
- good -goods, 1 Jn. 3:17
- goodman -my husband, Prov. 7:19 -master, Mt. 20:11; 24:43; Mk. 14:14; Lk. 12:39; 22:11
- governor -pilot, Jas. 3:4
- graffed -grafted, Rom. 11:17, 19, 23, 24
- grasshoppers -locusts, Amos 7:1
- grate -grating, Ex.27:4
- grief -sickness, Jer. 6:7
- grudge -grumble, Jas. 5:9; 1 Pet. 4:9 -tarry all night, Ps. 59:15
- grave -engrave. Ex. 28:9, 36; 2 Chr. 2:7
- graved -carved, 1 Kgs. 7:36
- graven -engraven, Ex.39:6
- guide -companion, Ps. 55:13
- guilty -bound. Mt. 23:18
- gutter -watercourse, 2 Sam. 5:8

- habergeon(s) -coat(s) of mail, Ex. 28:32; 39:23; 2 Chr. 26:14; Neh. 4:16; Job 41:26

- halted -limped, Gen. 32:31

- habitations -pastures, Amos 1:2

- hap -lot. Ruth 2:3

- hardly -with difficulty, Mt. 19:23

- hardly bestead -sore distressed. Is. 8:21

- harness -armor, 1 Kgs. 20:11; 22:34; 2 Chr. 9:24; 18:33

- harnessed -armed, Ex. 13:18

- hasted -hastened. Gen. 18:7; Job 31:5 -urged, Ex. 5:13

- hasten -watch over, Jer. 1:12

- hasteth -hasteneth, Job 40:23

- helve -handle, Deut. 19:5

- heresies -factions, 1 Cor. 11:19; Gal. 5:20; 2 Pet. 2:1

- heretic -factious. Tit. 3:10

- honorable -honored, Gen. 34:19

- horseleech -leech, Prov. 30:15

- hough(ed) -hock(ed), Josh. 11:6,9; 2 Sam. 8:4; 1 Chr. 18:4

- Huzzab -it is decreed, she, Nah. 2:7

- imagination -stubbornness, Deut. 29:19; Jer. 7:24; 9:14; 11:8; 13:10; 16:12; 18:12

- imagine -devise, Hos. 7:15

- influences, sweet -cluster, Job 38:31

- instant -insistent, Lk. 23:23 -constant, Rom. 12:12 -urgent, 2 Tim. 4:2

- instantly -diligently, Lk. 7:4 -earnestly, Acts 26:7

- instrument -weapon, Is. 54:16

- intermeddleth -quarrelleth, Prov. 18:1

- inward -familiar, Job 19:19

- inventions -doings, Ps. 99:8; 106:29,39

- isle(s) coast -land(s), Is. 20:6; 23:2; 24:15

- jeoparded -jeopardized, Judg. 5:18

- Jewry -Judea, Lk.23:5; Jn.7:1

- judgment -justice, Prov. 21:15; Amos 5:15,24; Mic. 3:9 -by reason of injustice, Prov. 13:23

- juniper roots -broom-roots, Job 30:4

- justle -jostle, Nah. 2:4

- keep under -buffet, 1 Cor. 9:27

- kindle -burn, Jer.33:18

- kine -cows. Gen. 32:15; 41:2, 3 (2 times), 4 (2 times), 18, 19, 20 (2 times), 26, 27; Deut. 7:13; 28:4,18,51; 32:14; 1 Sam. 6:7 (2 times), 10, 12, 14; 2 Sam. 17:29; Amos 4:1

- lade -load, Lk. 11:46

- lamb to -lambs for, Is. 16:1

- large -much. Mt. 28:12

- lay to -make. Is. 28:17

- leasing -lies. Ps. 5:6 -falsehood. Ps. 4:2

- length -last, Prov. 29:21

- lent -granted, Ex. 12:36

- let -loose. Ex.5:4 -hindered, Rom. 1:13 -restrain. 2 Thes. 2:7

- lewd -wicked, Acts 17:5

- lewdness -villainy, Acts 18:14

- light -lamp. 2 Kgs. 8:19 -vile, Num.21:5

- lightly -easily, Gen.26:10

- lightness -vain boasting, Jer. 23:32

- liketh -pleaseth, Deut.23:16; Amos 4:5

- liking, worse -worse looking, Dan. 1:10

- liquor -juice, Num. 6:3
- listed -would, Mt. 17:12; Mk. 9:13
- listeth -willeth, Jn. 3:8; Jas. 3:4
- lively -living, Acts 7:38; 1 Pet. 1:3; 2:5
- loft -the chamber, 1 Kgs. 17:19 -story, Acts 20:9
- Lucifer -Day-star, Is. 14:12
- lunatic -epileptic, Mt. 4:24; 17:15
- mad -foolish, Eccl. 7:7
- maid -virgin, Deut. 22:14,17; Job 31:1
- make away -sweep away, Dan. 11:44
- manner, with the -in the act, Num. 5:13
- marred -ruined, Mk.2:22
- maul -club, Prov.25:18
- mean -obscure, Prov. 22:29
- meat -meal, (multiple references too many to include here) -food, (multiple references too many to include here) -bread, Lev. 22:11, 13; 2 Sam. 13:5
- merchantman -merchant, Mt. 13:45
- mete -measure, Ex. 16:18; Ps. 60:6

- meteyard -measure of length, Lev. 19:35
- minished -diminished, Ps. 107:39
- minister -attendant, Lk. 4:20
- minstrels -flute players, Mt. 9:23
- moist -fresh, Num. 6:3
- month -new moon, Hos.5:7
- motions -passions, Rom. 7:5
- mount (of) -the hill country of Josh. 20:7; Judg. 2:9; 3:27; 4:5; 7:24; 10:1; 12:15; 17:1; 18:2; 19:1,16,18; 1 Sam. 1:1; 9:4; 14:22; 2 Sam. 20:21; 1 Kgs. 4:8; 12:25; Jer.4:15
- mount -mound, Ezek. 4:2
- munition(s) -stronghold, Is. 29:7; 33:16; Nab. 2:1
- naturally -truly, Phil. 2:20
- naughty -bad, Jer.24:2
- naughtiness -wickedness, Jas. 1:21
- nephews -grandchildren, 1 Tim. 5:4
- nether -lower, Deut. 24:6
- observed him -kept him safe. Mk. 6:20
- observed -kept, Gen. 37:11

- occupy (occupied) -trade(d), Lk. 19:13; Ezek.27:21
- occupied -used, Ex. 38:24
- occupied in -traded for, Ezek. 27:16, 19, 22
- occupiers -dealers in, Ezek. 27:27
- offend -be held guilty, Jer. 2:3
- oil olive -olive oil, Ex. 27:20; 30:24; Lev. 24:2; 2 Kgs. 18:32 -olive trees, Deut. 8:8
- open -frequent, 1 Sam. 3:1
- orator, eloquent -skillful enchanter, Is. 3:3
- organs -pipe, Ps. 150:4
- other (some) -others, Jn. 21:2; Acts 15:2; 17:18; 2 Cor. 13:2; Phil. 2:3
- ouches -settings. Ex. 28:11, 13, 14. 25; 39:6, 13, 16, 18
- outlandish -foreign. Neh. 13:26
- overcharge(d) -overburden(ed), Lk. 21:34; 2 Cor.2:5
- overlaid it -lay upon it. 1 Kgs. 3:19
- owl(s) -ostrich(es). Mic. 1:8; Deut. 14:15; Job 30:29; Is. 13:21; 34:13; 43:20; Jer. 50:39
- ox, wild -antelope. Deut. 14:15
- palace -turret, Song 8:9

- part -share. 1 Sam. 30:24 (3 times)
- particularly -in detail, Acts 21:19; Heb. 9:5
- passengers -to those passing, Prov. 9:15
- pastor(s) -ruler(s), Jer. 2:8 -shepherds, Jer.3:15; 10:21; 12:10; 17:16; 22:22; 23:1
- pate -crown. Ps. 7:16
- peep(ed) -chirp(ed). Is. 8:19; 10:14
- perish -be ruined. Mt. 9:17; Lk. 5:37
- pilled -peeled, Gen. 30:37,38
- pitiful -merciful. 1 Pet. 3:8
- very pitiful -full of pity, Jas. 5:11
- plain -quiet. Gen. 25:27
- plat -portion, 2 Kgs. 9:26 (2 times)
- platted -plaited, Mt. 27:29; Mk. 15:17; Jn. 19:2
- poll(ed) -cut the hair of, 2 Sam. 14:26 (3 times); Ezek. 44:20; Mic. 1:16
- polluted (polluting) -profaned (profaning), Is. 47:6; 56:2. 6
- pommels -bowls, 2 Chr. 4:12 (2 times), 13
- possessed -formed, Ps. 139:13

- pots -sheepfolds, Ps. 68:13

- power -striven, Gen. 32:28

- precious, more -rarer, Is. 13:12

- presently -straightway, Mt. 26:53; Phil. 2:23 -immediately, Mt. 21:19

- pressed out of -oppressed beyond, 2 Cor. 1:8

- presses -vats, Prov.3:10

- prevent(ed) -come before. Job 41:11; Ps. 88:13; Amos 9:10 -receive, Job 3:12 -precede, 1 Thes. 4:15 -spoke first to, Mt. 17:25 -met, Is. 21:14 -came upon, 2 Sam. 22:6, 19; Job 30:27; Ps. 18:5, 18 -anticipate(d), Ps. 119:147, 148

- prevent(est) -meet(est). Ps. 21:3; 59:10; 79:8

- printed -inscribed, Job 19:23

- privily -secretly, Ps. 10:8; 11:2

- profited -advanced, Gal. 1:14

- profiting -progress, 1 Tim. 4:15

- proper -own, Acts 1:19; 1 Cor. 7:7 -beautiful, Heb. 11:23

- prove(d) -test(ed), Dan. 1:12, 14

- purely -thoroughly, Is. 1:25

- purtenance -entrails, Ex. 12:9

- quick -alive, Lev. 13:10; Num. 16:30; Ps. 55:15; 124:3 - living, Heb.4:12

- quit -guiltless, Josh. 2:20

- quit you -acquit yourselves, 1 Cor. 16:13

- raised up -awakened, Song 8:5

- ranges -ranks, Lev. 11:35; 2 Kgs. 11:8, 15; 2 Chr. 23:14; Job 39:8

- ranging -roaming. Prov. 28:15

- reason -reasonable, Acts 6:2

- record -witness, Jn. 1:19; Acts 20:26; 2 Cor. 1:23; Phil. 1:8

- regard -preserve. Prov. 5:2

- relieve -refresh, Lam. 1:11,19

- removed, be -sway, Is. 24:20 -give way, Is. 22:25

- removed woman -woman in her impurity, Ezek.36:17

- removing -wandering, Is. 49:21

- rentest -rendest, Jer. 4:30

- repeateth -harpeth on. Prov. 17:9

- reprobate -refuse, Jer. 6:30

- reprove -decide. Is. 11:4

- residue -rest, Jer. 39:3; 41:10; Ezek. 9:8
- respect, had -looked, Heb. 11:26
- rest, take my -be silent, Is. 18:4
- rest resting -place, Ps. 132:8
- restrain -limit, Job 15:8
- ribband -cord, Num. 15:38
- right -steadfast, Ps. 51:10
- roaring -groaning, Ps. 22:1
- robbery -violence. Prov. 21:7
- rock -Sela, Is. 42:11
- rod -shoot, Is. 11:1
- roebuck -gazelle, Deut. 12:15, 22; 14:5; 15:22
- room(s) -stead, 1 Kgs. 5:1,5; 8:20; 20:24; 2 Kgs. 23:34; 1 Chr. 4:41; 2 Chr. 6:10 -in the room instead, 2 Sam. 19:13 -place. Mt. 2:22; Acts 24:27; 1 Cor. 14:16 place. Lk. 14:9,10 -seat, Lk. 14:8
- rooms -seats, Lk. 14:7
- sardine -sardius. Rev. 4:3
- satyr(s) -wild goat(s), Is. 13:21; 34:14
- savor -fragrance, Song 1:3

- scattered -tall, Is. 18:2,7 -broken, Ps. 60:2

- scrip -wallet, 1 Sam. 17:40 -bag, Mt. 10:10; Mk. 6:8; Lk. 9:3; 10:4; 22:35, 36

- secondarily -secondly, 1 Cor. 12:28

- secret -counsel, Prov. 3:32

- seethe -boil, Ex. 16:23 (2 times); 23:19; 29:31; 34:26; Deut. 14:21; 2 Kgs 4:38; Ezek. 24:5; Zech. 14:21

- seething -boiling. 1 Sam. 2:13

- senators -elders, Ps. 105:22

- sentence -judgment, Acts 15:19

- settle -ledge, Ezek. 43:14 (3 times), 17, 20; 45:19

- sever -distinguish, Ex. 9:4

- several -single, 2 Chr. 28:25; 31:19; -particular, Mt. 25:15

- shamefacedness -propriety. 1 Tim. 2:9

- shape -form. Jn.5:37

- shittah -acacia, Is. 41:19

- shittim -acacia, Ex. 25:5, 10, 13, 23, 28; 26:15, 26, 32, 37; 27:1, 6; 30:1, 5; 35:7, 24; 36:20,31,36; 37:1,4,10, 15,25,28; 38:1,6; Deut. 10:3

- shoot -pass, Ex. 36:33

- should -would. Acts 23:27
- show -tell, Gen. 46:31
- sides -innermost parts. 1 Sam. 24:3; Jon 1:5
- sincere -pure, 1 Pet. 2:2
- singular -hard, Lev. 27:2
- situate -situated, 1 Sam. 14:5
- skill, can -has skill (could skill-had skill), 1 Kgs. 5:6; 2 Chr. 2:7,8; 34:12
- smell -take no delight in, Amos 5:21
- snuffed up the wind -pant for air, Jer. 14:6
- sod -boiled, Gen. 25:29; 2 Chr. 35:13
- sodden -boiled, Ex. 12:9; Lev. 6:28 (2 times); Num. 6:19; 1 Sam. 2:15
- softly -gently, Gen. 33:14
- sore -fierce, 1 Sam. 14:52; 2 Sam. 2:17
- sorrowful -loathsome, Job 6:7
- sorrows -cords, Ps. 18:4,5
- speckled -sorrel, Zech. 1:8
- spirit -breath, Is. 40:7 -wind, Eccl. 11:5

- spoil, need of -lack of gain. Prov.31:11
- spoiled -plundered, Gen. 34:27,29
- spoiling -bereaving, Ps. 35:12
- spouse -bride, Song 4:8,9,10,11; 5:1
- stay -rely. Is. 30:12; 31:1; 50:10 -uphold, Prov. 28:17
- stem -stock. Is. 11:1
- stonesquarers -Gebalites, 1 Kgs. 5:18
- strain at -strain out, Mt. 23:24
- strait -narrow, 2 Kgs. 6:1
- straiten -distress, Jer. 19:9
- straitness -distress, Jer. 19:9
- strange -foreign, Acts 26:11
- strangers of -visitors from, Acts 2:10
- strength -rock, Is. 26:4 -stronghold, Ps. 31:4; Is. 23:14; 25:4 (2 times)
- strike -put. Ex. 12:7 -touch. Ex. 12:22
- string -bond, Mk, 7:35
- study -strive. 1 Thes. 4:11; 2 Tim. 2:15
- stuff -baggage, 1 Sam. 10:22; 25:13; 30:24

- substance -stock, Is. 6:13 (2 times)
- all the substance -every living thing, Deut. 11:6
- suburbs -the open space, Ezek. 45:2
- sycamore -sycomore, Amos 7:14
- Syrian -Aramaic, Is. 36:11
- table(s) -tablet(s), Lk. 1:63; 2 Cor. 3:3
- tablets -armlets, Ex. 35:22
- tabret(s) -timbrel(s), Gen. 31:27; 1 Sam. 10:5; 18:6; Job 17:6; Is. 5:12; 24:8; 30:32; Jer. 31:4; Ezek. 28:13
- taches -clasps, Ex. 26:6 (2 times), 11 (2 times), 33; 35:11; 36:13 (2 times), 18; 39:33
- take no thought -be not anxious, Mt. 6:25, 28,31,34; 10:19; Lk. 12:11,22,26
- take thought for -worry about, 1 Sam. 9:5; Mt. 6:27; Lk. 12:25
- tale -number, Ex. 5:8. 18; 1 Sam. 18:27; Chr. 9:28
- target -javelin, 1 Sam. 17:6
- teil tree -terebinth, Is. 6:13
- tell -count, Ps. 22:17
- told -counted, 2 Kgs. 12:10, 11; 2 Chr. 2:2

- temperance -self-control, Acts 24:25; Gal. 5:23; 2 Pet. 1:6
- temperate -self-controlled, 1 Cor. 9:25; Tit. 1:8
- tin -alloy, Is. 1:25
- Tirshatha -governor, Ezra 2:63; Neh. 7:65, 70; 8:9; 10:1
- tokens -signs, Ps. 135:9
- toward -from, Jer. 1:13
- translate(d) -transfer(red), 2 Sam. 3:10; Col. 1:13; Heb. 11:5
- traveleth, one that -a robber, Prov. 24:34
- trow -think, Lk. 17:9
- turning away -backsliding, Prov. 1:32
- turtle(s) -turtledove(s), Lev. 12:8; 15:29; Num. 6:10; Song 2:12; Jer. 8:7
- uncomely -unseemly, 1 Cor. 7:36
- unicorn(s) -wild ox (oxen). Num. 23:22; 24:8; Deut. 33:17; Job 39:9, 10; Ps. 22:21; 29:6; 92:10; Is. 34:7
- unperfect -unformed, Ps. 139:16
- upon -beside, Amos 9:1 -before, Nab. 3:5
- untoward -crooked. Acts 2:40

- usury -interest, Is. 24:2 (2 times); Mt. 25:27; Lk. 19:23
- vagabond -wandering, Acts 19:13
- vale -lowland, Deut. 1:7
- vain -evil. Jer. 4:14 -lying, Ex. 5:9
- vanity -falsehood, Ps. 12:2; 24:4; 41:6; Prov.30:8 -deceit, Ps. 144:8, 11 -destruction, Is. 30:28 -a breath, Is. 57:13 -wickedly. Is. 58:9
- vehement -sultry, Jon. 4:8
- vengeance -justice, Acts 28:4
- victory, in -forever, Is. 25:8
- virtue -power. Mk. 5:30; Lk. 6:19; 8:46
- void -an open, 1 Kgs. 22:10
- volume -roll. Heb. 10:7
- wanted -lacked, Jn. 2:3
- ware -wary, 2 Tim. 4:15 -aware, Acts 14:6
- waster -destroyer, Prov. 18:9
- wealth -good, 1 Cor. 10:24
- well -good, Mt. 12:12
- wench -maidservant, 2 Sam. 17:17

- whale -monster, Ezek. 32:2
- whole -healed, Josh. 5:8
- wine bottles -wine skins. Josh. 9:4 et al.
- wist -knew, Ex. 16:15; 34:29; Lev. 5:17, 18; Josh. 2:4; 8:14; Judg. 16:20; Mk. 9:6; 14:40; Lk. 2:49; Jn. 5:13; Acts 12:9; 23:5
- wit -know, Gen. 24:21; Ex. 2:4; 2 Cor. 8:1
- wot -know, Gen.21:26;44:15; Ex.32:1, 23; Num. 22:6; Josh. 2:5; Acts 3:17; 7:40; Rom. 11:2; Phil. 1:22
- wotteth -knoweth, Gen. 39:8
- woman -wife, Is. 54:6
- wood -forest. Is. 7:2
- work -recompense, Is. 40:10; 49:4; 62:11
- worship -honor. Lk. 14:10
- worthies -nobles, Nab. 2:5
- wounds -dainty morsels, Prov. 26:22
- wrap -weave together, Mic. 7:3
- wrung out to -drained by, Ps. 73:10
- yet -surely. Hos. 12:8

Ref- Holy Bible, Eng. Bible KJ53 Series, ABS-1973-150,000-D-7

It is plain to see that the changes in words and meanings in the KJV are beyond extensive. These revelations should make any bible scholar, and also Christian person, reconsider which version of the Bible that they turn to for guidance. Not to mention, the history of King James himself is one that is scattered with controversy. But that is another book all in itself. The only authority that should have a hand in writing, or revising the Holy Bible, is God Himself. How many versions of the Bible have passed through the hands of man and been manipulated for their selfish benefit? As we continue to review the Mandela Effect Bible changes, we shall consider the hand of man playing a huge role in these results.

The next Bible change that came across my path is by far the one that has shaken me to my core, along with thousands of others. Do you remember the Lord's Prayer? The most commonly spoken prayer at both funerals and Sunday's in church, is causing an enormous amount of controversy and confusion. When I spoke the Lord's prayer as a child, I always remember it saying, "Forgive us our trespasses," now it says, "Forgive us our debts." Since when did money or debt have anything to do with forgiveness? I will now share with you the various versions of the Lord's prayer as found today. Then you can decide what you remember.

The Lord's Prayer-

The prayer as it occurs in the ESV version of Matthew 6:9-13-

Our Father in heaven,

hallowed be your name.

Your kingdom come,

your will be done,

on earth as it is in heaven.

Give us this day our daily bread,

and forgive us our debts,

as we also have forgiven our debtors.

And lead us not into temptation,

but deliver us from evil.

The quite different form in the ESV version of Luke 11:2-4-

Father, hallowed be your name.

Your kingdom come.

Give us each day our daily bread,

and forgive us our sins,

for we ourselves forgive everyone who is indebted to us.

And lead us not into temptation.

Original Greek and translated Latin versions-

Nestle Aland 28:

Πάτερ ἡμῶν ὁ ἐν τοῖς οὐρανοῖς·

ἁγιασθήτω τὸ ὄνομά σου·

ἐλθέτω ἡ βασιλεία σου·

γενηθήτω τὸ θέλημά σου,·

ὡς ἐν οὐρανῷ καὶ ἐπὶ γῆς·

τὸν ἄρτον ἡμῶν τὸν ἐπιούσιον δὸς ἡμῖν σήμερον·

καὶ ἄφες ἡμῖν τὰ ὀφειλήματα ἡμῶν,

ὡς καὶ ἡμεῖς ἀφήκαμεν τοῖς ὀφειλέταις ἡμῶν·

καὶ μὴ εἰσενέγκῃς ἡμᾶς εἰς πειρασμόν,

ἀλλὰ ῥῦσαι ἡμᾶς ἀπὸ τοῦ πονηροῦ.

Translated Latin Version Nova Vulgata-

Pater noster, qui es in caelis,

sanctificetur nomen tuum,

adveniat regnum tuum,

fiat voluntas tua,

sicut in caelo, et in terra.

Panem nostrum supersubstantialem da nobis hodie;

et dimitte nobis debita nostra,

sicut et nos dimittimus debitoribus nostris;

et ne inducas nos in tentationem;

sed libera nos a Malo.

1662 Anglican BCP-

Our Father, which art in heaven,

hallowed be thy name;

thy kingdom come;

thy will be done,

in earth as it is in heaven.

Give us this day our daily bread.

And forgive us our trespasses,

as we forgive them that trespass against us.

And lead us not into temptation;

but deliver us from evil.

When before the Collect the priest alone recites the prayer, the people here respond: Amen.

When after all have communicated the people repeat each petition after the priest, the prayer ends:

For thine is the kingdom,

the power, and the glory,

for ever and ever.

Amen.

Matthew 6:9-13 (ESV)-

"Pray then like this: 'Our Father in heaven, hallowed be your name. Your kingdom come, your will be done, on earth as it is in heaven. Give us this day our daily bread, and forgive us our debts, as we also have forgiven our debtors. And lead us not into temptation, but deliver us from evil.'"

Luke 11:2-4 (ESV)-

And he said to them, "When you pray, say: 'Father, hallowed be your name. Your kingdom come. Give us each day our daily bread, and forgive us our sins, for we ourselves forgive everyone who is indebted to us. And lead us not into temptation.'"

Matthew 6:9-13King James Version (KJV)-

9 After this manner therefore pray ye: Our Father which art in heaven, Hallowed be thy name.

10 Thy kingdom come, Thy will be done in earth, as it is in heaven.

11 Give us this day our daily bread.

12 And forgive us our debts, as we forgive our debtors.

13 And lead us not into temptation, but deliver us from evil: For thine is the kingdom, and the power, and the glory, forever. **Amen.**

Now that we have reviewed the various translations of the Lord's Prayer that exists today, are you shocked? Do you remember the KJV always saying "debts" or do you remember "trespasses"? Do you remember us being *on* Earth or *in* Earth as it is in Heaven? There is still a huge debate rolling around the Mandela Effect community regarding these changes, so I will now continue to share the other changes that have arisen out of the Mandela Effect.

One of the most famous images in Christian culture is without a doubt, the Lion and the Lamb. The lion symbolizes strength, courage, and leadership. A lion's roar stops you; it is a terror that shakes you up and draws you into something beautiful. The Lamb of God (Greek: Ἀμνὸς τοῦ Θεοῦ, Amnos tou Theou; Latin: Agnus Deī [ˈaŋ.nʊs ˈde.iː]) is a title for Jesus that appears in the Gospel of John. It appears at John 1:29, where John the Baptist sees Jesus and exclaims, "Behold the Lamb of God who takes away the sin of the world." We often see these two animals side by side. However, the next Mandela Effect Bible change shifts this image remarkably.

The Lion and the Lamb
Author- images4.fanpop.com/

If you conduct an Internet search today and ask the question: What is the meaning of the lion and the lamb? You receive the Bible scripture known as Isaiah 11:6. But now the scripture does not exclaim that the Lion lies down with the Lamb anymore, it now lays down with the WOLF. If you recall this scripture differently, then you will be shocked when you read how it is today:

> Isaiah 11:6: The **wolf** also shall dwell with the **lamb**, and the leopard shall lie down with the kid; and the calf and the young lion and the fatling together, and a little child shall lead them. It's talking about the good days to come when the Jews return to Israel, and the Messiah comes.

I thought that it would be important to see what other Christian scholars had to say about this passage, so I referenced a popular website known as GotQuestions.org. Here are the credentials and mission statement of Got Questions Ministries, so that we can confirm their validity in their expertise:

"Got Questions Ministries seeks to glorify the Lord Jesus Christ by providing biblical, applicable, and timely answers to spiritually-related questions through an internet presence."

GotQuestions.org is a ministry of dedicated and trained servants who have a desire to assist others in their understanding of God, Scripture, salvation, and other spiritual topics. We are Christian, Protestant, conservative, evangelical, fundamental, and non-denominational. We view ourselves as a para-church ministry, coming alongside the church to help people find answers to their spiritually related questions.

We will do our best to prayerfully and thoroughly research your question and answer it in a biblically-based manner. It is not our purpose to make you agree with us, but rather to point you to what the Bible says concerning your question. You can be assured that your question will be answered by a trained and dedicated Christian who loves the Lord and desires to assist you in your walk with Him. Our writing staff includes pastors, youth pastors, missionaries, biblical counselors, Bible/Christian college students, seminary students, and lay students of God's Word.

All of our answers are reviewed for biblical and theological accuracy by our staff. Our CEO, S. Michael Houdmann, is ultimately accountable for our content, and therefore maintains an active role in the review process. He possesses a Bachelor's degree in Biblical Studies from Calvary University and a Master's degree in Christian Theology from Calvary Theological Seminary (Kansas City, MO).

May God richly bless you as you seek to study His Word and grow in your walk with Him! (Romans 11:36)." www.gotquestions.org

On their website, I was able to locate a section of their site dedicated to the Lion and the Lamb symbolism, I think it offers an alternative view to this scripture debate, but in the end, it will be up to *you* to decide.

Question: "How should we understand the Lion and the Lamb passage?"

Answer: Typically, when someone is thinking of the "lion and the lamb," Isaiah 11:6 is in mind due to it often being misquoted, "And the wolf will dwell with the lamb, and the leopard will lie down with the young goat, and the calf and the young lion and the fatling together." The real "Lion and the Lamb" passage is Revelation 5:5-6. The Lion and the Lamb both refer to Jesus Christ. He is both the conquering Lion of the tribe of Judah and the Lamb who was slain. The Lion and the Lamb are descriptions of two aspects of the nature of Christ. As the Lion of Judah, He fulfills the prophecy of Genesis 49:9 and is the Messiah who would come from the tribe of Judah. As the Lamb of God, He is the perfect and ultimate sacrifice for sin.

The scene of Revelation 4—5 is the heavenly throne room. After receiving the command to write to the seven churches in Asia Minor, John is "caught up in the spirit" to the throne room in heaven where he is to receive a series of visions that culminate in the ultimate victory of Christ at the end of the age. Revelation 4 shows us the endless praise that God receives from the angels and the 24 elders. Chapter 5 begins with John noticing that there is a scroll in the "right hand of him who was seated on the throne." The scroll has writing on the inside and is sealed with seven seals.

After giving us a description of the scroll, an angel proclaims with a loud voice, "Who is worthy to open the scroll and break its seals?" John begins to despair when no one comes forth to answer the angel's challenge. One of the 24 elders encourages John to "weep no more," and points out that the Lion of the tribe of Judah has come to take and open the scroll. The Lion of the tribe of Judah is obviously a reference to Christ. The image of the lion is meant to convey kingship. Jesus is worthy to receive and open the scroll because he is the King of God's people.

Back in Genesis 49:9, when Jacob was blessing his sons, Judah is referred to as a "lion's cub," and in verse 10 we learn that the "scepter shall not depart from Judah." The scepter is a symbol of lordship and power. This was a prophecy that in Israel the kingly line would be descended from Judah. That prophecy was fulfilled when David succeeded to the throne after the death of King Saul (2 Samuel). David was descended from the line of Judah, and his descendants were the kings of Israel/Judah until the time of the Babylonian captivity in 586 BC.

This imagery of kingship is further enhanced when Jesus is described as the "root of David." This harkens us back to the words of Isaiah the prophet: "There shall come forth a shoot from the stump of Jesse, and a branch from his roots shall bear fruit. . . . In that day the root of Jesse, who shall stand as a signal for the peoples—of him shall the nations inquire, and his resting place shall be glorious" (Isaiah 11:1, 10). As the root of David, Jesus is not only being identified as a descendant of David, but also the source or "root" of David's kingly power.

Why is Jesus worthy to open the scroll? He is worthy because He "has conquered." We know that, when Jesus returns, He will conquer all of God's enemies, as graphically described in Revelation 19. However, more importantly, Jesus is worthy because He has conquered sin and death at the cross. The cross was the ultimate victory of God over the forces of sin and evil. The events that occur at the return of

Christ are the "mop-up" job to finish what was started at the cross. Because Jesus secured the ultimate victory at Calvary, He is worthy to receive and open the scroll, which contains the righteous judgment of God.

Christ's victory at the cross is symbolized by his appearance as a "Lamb standing, as though it had been slain" (Revelation 5:6). Before the exodus from Egypt, the Israelites were commanded by God to take an unblemished lamb, slay it, and smear its blood on the doorposts of their homes (Exodus 12:1-7). The blood of the slain lamb would set apart the people of Israel from the people of Egypt when the death angel came during the night to slay the firstborn of the land. Those who had the blood of the lamb would be spared. Fast forward to the days of John the Baptist. When he sees Jesus approaching him, he declares to all present, "Behold, the Lamb of God, who takes away the sin of the world!" (John 1:29). Jesus is the ultimate "Passover lamb" who saves His people from eternal death.

So when Jesus is referred to as the Lion and the Lamb, we are to see Him as not only the conquering King who will slay the enemies of God at His return but also as the sacrificial Lamb who took away the reproach of sin from His people so they may share in His ultimate victory.

Retrieved from: https://www.gotquestions.org/Lion-and-the-Lamb.html

Got questions certainly raises an interesting debate. However, I find that it can be difficult to trust the Internet as the final word of anything, as I know it can also be manipulated. I also do not know if these sites existed before the Mandela Effect or not, so I have to wonder if there has been some covering of tracks at play here. I now leave this up to your interpretation, what do you think the Lion and Lamb represents?

Has the wolf always been in the scripture for you?

After contacting Byron Preston (Harmony Mandela Effect) regarding his research on the Mandela Effect, I was pleasantly surprised to find out that Byron had also received a Theological degree. I had no idea that he had not only received an education in physics, but also in the word of God. Upon making this discovery, Byron explained to me that he had also recorded a video on the subject of the Wolf and the Lamb, he has given me permission also to share his work on this topic.

Wolf and the Lamb EXPLAINED!:
https://youtu.be/IYfwJWIdVt4

At the start of the video, Byron gives his educational background within the Church. He then goes on to explain that most people consider the Holy Bible to be the infallible word of God, yet there is some confusion when we know that this word has frequently passed through the hands of men. The men who penned the Bible had all claimed to be inspired by the Holy Spirit to carry out their works. But what does the word "inspired" truly mean? Inspired: of extraordinary quality, as if arising from some external creative impulse, or (of air or another substance) that is breathed in. At no point in this definition, do we see that an inspired word is infallible? At the end of the day, the Bible was written by the hands of men, and men can often make mistakes.

Knowing this fact to be true, we have to be very careful how much meaning we place on the pages of these books. Like I have repeated time and time again, the word of God only lives within your heart. His message is written on your soul, and that meaning cannot change. If we can keep this in mind when reviewing Mandela Effect Bible changes, we will not allow ourselves to be overcome by fear. As a matter of fact, it gives me, even more, faith as I know that His word must be mighty, or "they" would not be inspired to change it. If they are changing His word, they must be threatened by it.

Byron believes that there is some level of changes happening to the Holy Bible to awaken many Christians from their sleep. A majority of Christian people today believe in what many refer to as the rapture. They believe that Armageddon must take place for Jesus Christ to return and save us all. But is this a passive behavior to only wait for His return without doing any of the work themselves? We must take accountability for our roles in life; we cannot just sit back and wait for someone to save us. This attitude and belief system is what some people call lukewarm Christians, they read the books, go to church, but at the end of the day are just waiting to be rescued.

Several of the Bible changes recorded with the Mandela Effect, all seem to point to the concept of singularity. In essence, the changes we see, are adjusting in a way that we can understand that we are all one, we are all connected to the same God, and that we are all His children. Whether we speak about a wolf or a lamb, both of these creatures were formed by God. There is no evil in God's creations; we are all born into perfection in His eyes.

Byron also goes on to express that when he presents Bible changes to righteous Christians, they immediately respond with judgment. Our God is a loving and forgiving Creator, yet so many people want to label Him as judgmental and unforgiving, this is a fallacy. The most important trait that people can focus on as Christians, are to remove all judgment from their belief systems. I will expand upon the topic of judgment in the next section of Bible Changes. But for now, I will conclude this article with a collection of the public comments from Byron's video. I was overwhelmingly inspired by some of the comments that I read on his channel:

Public Comments on the Wolf and the Lamb:

Susie-Marie Woods-

I agree with your interpretation of the Bible.....a book that was inspired by imperfect humans who were inspired by the perfect God. Therefore, it is flawed. However, due to its inspiration, it is not to be put aside. There are good things in it. But, it needs to be understood according to what it is. I also do not believe in original sin. How can we be responsible for the actions of others? That doesn't make any sense. Wolf and lamb!? I grew up in a highly religious family and I definitely remember it being the lion and the lamb.

SRPHD-

awsesome video bro respekt brotha . sorry late reply... life lol. keep up the awsesome work brotha. respekt for props. ill mirror later on today brotha just reuping vids before i make my next 2 on consciousness. need ppl to see the others first or indoctrination will stop the understanding haha . respekt love and light brotha.

Midnight Heston-

I remember a song from Elvis Presley, "Peace in the Valley" The words are "The Bear will be gentle, the Wolf will be tame, and the Lion shall lay down with the Lamb."

Da.Nobody Nameless-

I saw the Bible be rewritten right in front of my eyes.

Anungquay10134-

I agree with the point that we are not born into sin. We are born perfect and we learn and choose to love or not. A loving environment fosters a loving human being. Unfortunately, our present environment leads many astray... and many are left neglected and angry. Love is the only answer.

BRANDON x Talks-

The Bibles has changed supernaturally, but only little words have changed... for example "bottles" has replaced "wine-skins." Mostly the King James Version has changed supernaturally. Yes it is a little change that doesn't really effect the parables or the meaning, but the fact is that it has changed miraculously. Know that God is allowing the changes, so it's ok. The Lord Jesus will "never forsake you." This miracle is real. The delusion is believing that this "miracle" has not happened. It's ok that it happened. God is still in control. Jesus is still Lord; He is my LORD. Just because the devil can do miracles, doesn't mean God is not in control. It is NOT nonsense. It is NOT bad memory.

The Word of God has not changed because the Bible is not the Word of God. It is Scripture (a word). Yes, it is true, but the Word is living. Jesus is the Word. The Word existed before the world, before the universe. The Word will never change, but little words in the Bible have changed miraculously. But just because words in the Bible changed, doesn't mean the Word is changed. The Word is within you. Don't worry that miracles are happening. Don't be deluded into turning your head away from the effects like they are not real- the Mandela Effects are real changes to reality but don't be afraid of it. God is still in control Jesus is still Lord. He is still my Lord, and everything will still be ok. Just be alert, more now than ever, because miracles are changing little meanings for future Christians who will be reading edited scriptures that they think have always existed that way, but in fact, we know that they have changed.

The staff of Moses turned into a snake, and that snake ate the pharaoh's witchcraft snakes. The devil wants to put fear in you, but don't be afraid of seeing miracles. Jesus conquered the world already, don't worry, even though the changes are real, doesn't mean your faith is wrong. Continue to trust the Word within you. The True Word only exists within; It is not the bible, the bible reflects the Word within. Trust God, Trust the Holy Spirit- for He lives within YOU.

Amor Fati-

I have been under the impression that the "infallible WORD of GOD" was Jesus since he is referred to as "the Word of God" since Genesis. He, being God, is the only thing capable of being infallible.... not the Bible. It was hard for me to understand and accept that "the Word of God" referenced the Bible when it's been historically tampered with, namely the canonization, and numerous times since then. When I finally heard a teaching over the original Hebrew vocabulary and definitions, going all the way back to creation, referencing "the Word" to Jesus and "the Wind" to the Holy Spirit. That was the first time I could feel my soul at rest with the understanding and meaning of infallibility. Is this your understanding as well, or have you heard the same or similar teachings?

Adam Colbert-

I tried convincing a "hardcore Christian" friend of mine that since we are to LOVE our enemies [Jesus's own words!], and that since Satan is the greatest enemy [it literally means "enemy" and "adversary"], then we are literally called to LOVE Satan. I explained that to "love" is NOT to "like." Love is an outward expression, radiance. Liking something is being similar to something, taking it IN. Also, to love someone does NOT necessarily mean to "enjoy" them or what they do. I gave scriptural evidence, the Beatitudes [edit: sermon on the mount] says that God's love is like the rain and the sun fall on / shines on ALL, both "good" plants and the weeds alike. Long story short, he wasn't able to accept it. He relied HEAVILY on some Old Testament (or rather "Old COVENANT") verses, speaking about how "God hates ____" and saying that we should follow that example. I know he means best and has genuinely striven to a godly man for as long as I know him, but he seems to have a "hate addiction." He just loves (or shall I say LIKES and ENJOYS) to hate the devil. Good Lord.....since WHEN did feeding an enemy, which thrives upon hate, WITH hate ever diminish said, enemy? And he wonders why they devil is following him everywhere??? I'm thinking "HELLO?!? BECAUSE YOU'RE *FEEDING* IT/HIM/WHATEVER!!!" Love conquers all. Love always wins. Love your enemy. LOVE your enemy! Not like, not enjoy, certainly not hate, but LOVE!!!!!!! :) <3

Ms.Rhythm Jones-

I never understood that scripture as Most of the Bible takes place in the middle east, I did not know there were wolves in the middle east...

Anonymous Anonymous-

I remember the Lion and the Lamb. Thanks for that info. I woke up three years ago and been like a rudderless boat for that whole time until you came along. It's amazing.

OldDood-

We also Reincarnate...That is how we learn and return to the Source...

There are certainly some various points to consider when reviewing Bible changes related to the Mandela Effect. After my extensive research into these changes, I am convinced that God,

Himself most certainly has a hand in what is happening with the Mandela Effect.

Do you believe the changes are supernatural in origin?

"Judge not, lest ye be judged..."

We have all heard this infamous saying, right? It tells us that it is important not to pass judgment upon another human being unwitting. As a Christian, I have often quoted this phrase when in the presence of a judgemental person. But is this the correct scripture to recite? Or have we all gotten *this* one wrong as well? Or has someone been editing and tampering with our Holy Bible's yet again?

Let us review this scripture as it is today:

Matthew 7:1-3 King James Version (KJV)

7 Judge not, that ye be not judged.

2 For with what judgment ye judge, ye shall be judged: and with what measure ye mete, it shall be measured to you again.

3 And why beholdest thou the mote that is in thy brother's eye, but considerest not the beam that is in thine own eye?

Judge not that ye be *not* judged? Am I reading this correctly? Have my eyes deceived me? Does that version of scripture not mean the exact opposite now? I am honestly perplexed along with countless other Mandela Effected individuals. Now I must see wherein this debate began.

I discovered the website of a man named Wayne Stiles. On his site, I located an informative article with his personal take on the judgment debate:

JUDGE NOT, LEST YE BE JUDGED—WHAT JESUS MEANT

Thankfully, He told us what He meant, so we don't have to guess.

-Wayne Stiles

The best-known Bible verse used to be John 3:16. But our culture has a new favorite. In fact, it has become the trump card played to justify any and every lifestyle. It's even a quote from Jesus.

"Judge not, that ye be not judged."

The phrase is often quoted as "Judge not, lest ye be judged." While the meaning is the same, it's interesting we have learned the wrong wording from the 1611 King James Version. It should be: "Judge not, that ye be not judged." The verse often is taken to mean nobody has the right to judge anybody for anything at any time.

The problem? The verse has a context. Jesus told us what He meant.

When Jesus spoke these words on the slopes surrounding the Sea of Galilee, He wasn't saying never to judge. He simply warned about doing it the wrong way—by telling us how to make judgments the right way.

And believe me, it ain't easy.

JESUS ISN'T CONFUSED

Later in the same book, Jesus commands we do confront a fellow Christian caught in a sin (Matthew 18:15-17). This awkward obligation is supported elsewhere in the Bible (Galatians 6:1).

So, what did Jesus mean, then, when He said, "Judge not lest ye be judged"? The verse that follows explains—and often it isn't quoted. Jesus tells us exactly what He meant:

Do not judge lest you be judged. For in the way you judge, you will be judged; and by your standard of measure, it will be measured to you. — Matthew 7:1-2

Jesus wasn't confused in His teaching. He didn't mean we should never make a judgment about right and wrong. As He explained, He meant we shouldn't make a judgment hypocritically. The verses that follow make this patently clear (Matthew 7:3-5).

PLEASE PASS THE PERFECTION

We know no one is perfect, but we expect it anyway. Except in ourselves. We often excuse our own shortcomings because we claim God's grace. But then we turn around and demand others be perfect—a standard we ourselves don't meet.

This is precisely what Jesus was warning against.

The fact is, we never know all the facts.

How do we know the idiot driver didn't just lose his spouse last week?

How do we know the rude saleswoman didn't just discover she has cancer?

How do we know the Christian who cussed didn't just accept Jesus and has no clue how to walk with God?

Wouldn't it be better to tap the brakes on our judgment—especially when we don't know all the facts? Before we call into question someone else's walk with God, we should scrutinize our own.

NOTE: the Bible never gives us the responsibility of pointing out sins of the world as our priority. Unbelievers don't need to hear lessons of morality as much as they need to hear the gospel. The judgment Jesus spoke of was primarily between believers, not unbelievers.

The world needs the gospel, not rules alone. Otherwise, they may confuse rules with the way of salvation.

JUDGE NOT, LEST YE BE JUDGED —WHAT JESUS MEANT

Yes, there is a point when we must confront the sin in another Christian's life. Otherwise, we're failing to obey the process Jesus and Paul explained in Matthew 18:15-17; Galatians 6:1; and 1 Corinthians 5:3-5.

However, that process should only occur after we've gone through a more basic examination with our own lives.

"Judge not lest ye be judged." Jesus meant that our priority for life change should first be to ourselves—then to others.

-Wayne Stiles

https://www.waynestiles.com/judge-not-lest-ye-be-judged-what-jesus-meant/

Wayne Stiles: A BRIEF BIO

I have served in full-time Christian ministry since 1991. After graduating from the University of North Texas and Dallas Theological Seminary, I served in the pastorate for 14 years at Denton Community Church, a church plant of Denton Bible Church. In 1995, I returned to seminary to complete my Master of Theology and Doctor of Ministry degrees.

While I appreciate what Wayne Stiles has attempted to present in this article, I could not help but feel the underlying tones of judgment directed at the person reading the article. I sometimes wonder of some of these "experts" existed before the Mandela Effect, as I have mentioned before. It seems as though every single time I conduct research of a supportive nature to my theory, I land on nothing but opposing arguments. This is a theme with the Mandela Effect; it always keeps you guessing. I will not doubt the validity of Mr. Stiles work, nor question his faith in God; I just have to wonder if he comes from a universe that thousands of others do not. I will now share with you some of the comments I received on my first YouTube video about Mandela Effect Bible changes:

Public Comments on Bible Changes:

Rich Rockwell-

I am 63 years old and remember the things you mentioned just like you do. The Bible scripture of the lion and the lamb really threw me a curve. Thanks for your video, people can get rather mean when you try to explain this effect. Thank You!

blessed777favored-

I LOVE YOU, Listen to My words. A Prophetic word just for you RIGHT NOW, at this time!! Saith the Lord. Wake up those that My Father has given unto Me, and hear my words, for I proclaim this truth unto you this day. He who has ears, let them hear, what I am saying to the true Church, which are those that know My Name, saith the Lord. Something that has been hidden from you from the beginning, along with many other truths, but I tell you this truth. I said that I would not do anything until I reveal it to my prophets/prophetess and that is what I am doing and have been doing through the words of My prophetess, saith the Lord, fulfilling the Father scripture. Amos 3:7 Surely the Lord GOD will do nothing, but he revealeth his secret unto his servants the prophets. I have not taught them every aspect of the Holy Scripture, saith the Lord because now they are changing before your eyes. So don't challenge them, the teaching is over. I have taught them the things that I wanted them to say when it is time to tell them. They are those that know My name, as well as the body that I am returning to you in your time. Something that was purposely kept from you so that you could run to the first one that is coming which has nothing in Me, saith the Lord, fulfilling the Father's scripture in.. John 5:43 John 14:30 Hereafter I will not talk much with you: for the prince of this world cometh, and hath nothing in me.

Stanley Plock-

The Lord's Prayer and other stuff to ponder... The Lord's Prayer used to say "Forgive us our TRESPASSES. Now it has been changed to: "Forgive us our DEBTS"? The Lord's Prayer has been modified from a GENERAL sin to a VERY SPECIFIC SIN that most people don't even know about. Creating money from NOTHING and charging INTEREST is a SCAM. This is the BIGGEST SIN AGAINST MANKIND IN ALL OF HISTORY! The Quantum Computer is to the OUIJA Board... As the Electronic Calculator is to the Abacus.

Sue Johnson-

The Lord's prayer to my Catholic cousins always said "trespasses" My sister and I learned forgive us our debts, and we belonged to an Evangelical and Reformed United Church of Christ. So...NO Mandela Effect, just different wording in different editions of the Bible.

steven lee-

Sister check out "2 Corinthians 11:8 all so" Luke 5:4 the word launch in the King James" even in most Bible's" oh really" the right word should be (CAST)" there is so much to tell "!

DeeaXiis-

I am a believer, but I have always known the Bible has been tampered with. Also, I think 911 was an inside job. I see the changes in Bible. I see the Effect in other things also, like the Mona Lisa. Capricorn was a goat but changed a few years ago to a MerGoat. I remember the change because I drew mermaids. Goat = Saturn = Satan. Also before I use to big on Astrology. I am a Sagittarius Dragon. I was the only mythical creature in the western and eastern zodiac,) ... Capricorn was not a mythical creature. Thank you for your video.

Roudy One-

I believe you the lord's prayer 100%, and it is trespasses and always was. And this thing will overwhelm you!

Mark Bendavid-

I'm shocked at how multiple changes actually conCERN 9:11, i.e.,., 1) Isaiah 11:6 (9:11 inverted) - WOLF 2) Comcast Commercials - also flaunts WOLF, build it HE will come, etc. all in like 30second Olympics commercial! 3) E-Team & Gelatin Group Israeli Art Students - Makes me wonder where the next 911, false flag, etc. will be?

mandelaeffects.com/forums/topic/gelatineteam
http://mysticjourneybookstore.com/and-the-lion-shall-lay-down-with-the-lamb-isaiah-11-11/

Life Matrix-

John of the Divine? What the Heaven? Never even seen Divine in the Bible.

art H-

I have two bibles close at hand the KJ and the New American Standard; the KJ version did change but, the NAS still says the Revelation to John.

Heart Mind-

Stasha, thank you for spreading this message in a positive and truthful light. I believe you are walking and fulfilling your mission. I've been telling people that we all have these different metaphysical abilities. That everything around our environment (fluoride) does suppress us from these natural abilities. And with practice and detox, we can achieve much more than previously known. I feel like I've been a lukewarm Christian for a while now. But with the recent events in my life without any doubt in my heart now, I know God and Jesus are Love and the true path to walk. Whether it's divine or man-made, there is definitely something occurring. And it's important for us to stay awake and to share our truth. Unity to ascend. I love you, and your energy hope you have a wonderful day. P.s. I even made a video the other day and wrote down Revelation 22:13 instead writing down Revelations. Which I was second-guessing myself because it didn't look right. I even spent all of 2012 pretty much just studying the last book of the Bible because I felt it was important. Also, I sent you an email earlier, hope you received it.

Salted Seeder-

Might I suggest you buy an NKJV Bible because it reads like we remember. Still has the revelation of Jesus Christ, wineskins, denarius, possessions, etc. Also, the KJV dedication to Queen Elizabeth is Queen Elizabeth Tudor the 1st dedicated from King James. Hopefully, you know about Constantine who is responsible for the KJV. I too remember revelation(s) but not positive I wasn't pluralizing it. I do like your angle, and you're definitely <- on my sub list. I was the one saying this is 100% God who works on a quantum level and that I see the clues and messages in the changes, the awakening from our false force-fed reality from the enemy. I do believe the enemy is trying to counteract but as we know God is in control and is always many steps ahead. I'm not doing a very good job at the moment explaining my theory, I'm all over the place but possibly in another comment, I'll go into detail, and it might be helpful to you or somebody else. We should all work together on this. Great video

Yarikk734-

Small point really but Constantine died probably 1200-1300 years before the KJV of the Bible was translated from the original languages into English. He was responsible/involved with the Council of Nicea however.

Roudy One-

For god sakes the Bible changes every day it seems like :/ crazy just scary and makes, you wonder what is next ???

Bride of Christ-

The Vatican has been omitting to the KJV Cambridge 1611 with all the added versions. The Vatican mind psy-op is to discredit the one Holy Bible, that which they never wanted released to the English-speaking world, so that's why the Vatican papacy put out the KJV Oxford edition in order to discredit the Cambridge 1611 King James Holy Bible. Fiona Broome (Nice name for a ghost writer), is taking credit for the ME, she is a Vatican troll.

Snow white first calls the mirror magic. THE Context of Isaiah 11:6 scripture, is a child will be safe, as the child leads otherwise known furious/vicious animals. We will go back to the garden where NO more death occurs. Also NOTE VERSE 9 Isaiah 11:6-9 King James Version (KJV) 6 The wolf also shall dwell with the lamb, and the leopard shall lie down with the kid; and the calf and the young lion and the fatling together; and *a little child shall lead them. 7 And the cow and the bear shall feed; their young ones shall lie down together: and the lion shall eat straw like the ox. 8 And the sucking child shall play on the hole of the asp, and the weaned child shall put his hand on the cockatrice' den. 9 They shall not hurt nor destroy in all my holy mountain: for the earth shall be full of the knowledge of the Lord, as the waters cover the sea.

gyoergy barabass-

Ellis Island was the island where new refugees arrived in NYC or????? The Bible was false several times!!! No Statue of Liberty on Ellis Island it was definitely loony toons not tunes - definitely.

Spiritual Freedom!!!-

I was searching for Mandela effects on the Bible and searched for Matthew Ch.6. What I got on the top of the search pages were the first three words of Matthew chapter 6, and then the word "Bible"... So, it reads "Be careful not Bible"... Tripped me out!! A sign that the bible really has changed? Or a coincidence? Maybe others can do this search too and tell me what you think!

Cori Castello-

I have been watching your Mandela videos, and other videos and reading about this topic for a long while. Everything you remember is what I remember. My boyfriend is on the fence about it. I almost feel like I am going crazy. But for years I have felt that I was having horrible memory issues, and I am glad to know that isn't the issue. It is scary though.

Roudy One-

I believe you the lord's prayer 100%, and it is trespasses and always was, and this thing will overwhelm you!

Paul Adams-

I don't know what led me to your channel, but you REALLY need to get a hold of me. I do have a channel on YouTube.... search Paul Adams and look for the long-haired guy.... lol. I also have created two Facebook groups based on these changes. Our groups have become large, and we also have a website. You are SO on the right path. I really would like to talk to you...

Diggy d-

I find the Christian values connection interesting because in my observation it does apply. While I was born in a Christian nation, I don't claim such. I was educated at a Christian school, a Catholic school, and finally Episcopalian. I "accepted" Jesus in my heart when I was a child... Looking back, I didn't mean it. How does a child have the ability to understand that? I just knew that something like that was the answer. The indoctrination I received really pushed God out of my life in my 20s. Even the times I was screaming at God for what an A-hole he was, I knew he was there. And even today I would never claim to be a Christian because I met so many fake shallow humans that I don't want to be associated with. I had a spiritual awakening about four years ago that changed my life. While I don't subscribe to any organized religion because you have God in you and you yourself are a god, I would say I have Christian values. Something is real. Whatever God is its love. That's all I felt. Love. It's interesting to me who sees the ME and doesn't. I've become weary of those that don't see it.

steven lee-

Sister check out "2 Corinthians 11:8 all so" Luke 5:4 the word launch in the King James" even in most Bible's" oh really" the right word should be (CAST)" there is so much to tell "!

Sue Johnson-

The Lord's prayer to my Catholic cousins always said "trespasses" My sister and I learned forgive us our debts, and we belonged to an Evangelical and Reformed United Church of Christ. So...NO Mandela effect. Just different wording in different editions of the Bible.

Patricia W-

Doesn't the new verse seem shorter?

Joan Clayton-

Nope, I remember it trespasses and trespassers. I said the Lord's Prayer daily sometimes 2 or 3 times a day for over two years in each AA meeting I attended back in 1995-1997. We all said it that way. We are not crazy!

Tappedline-

FYI, I am not a KJV guy... I like the Geneva 1599. Go back and read the Lord's Prayer and then read in the KJV... Fill in the Blank "as it is (blank) earth" I think you said "on" but is that what is written now? Peace.

Were Puppy-

The Isaiah 11:6 is driving me crazy.

TheVineRhyme-

Another important Mandela effect in scripture is found in Luke 17:31 where it now reads "In that day; he which shall be upon the housetop, and his "STUFF" in the house, let him not come down to take it away...: It used to say: "in that day, he that is on the housetop, let him not go down to take anything out of his house..." the word "STUFF" is crazy, since this is in the Original King James translation of 1611 and in 1611 the word "STUFF" wasn't even part of old English language for any sort of reference to possessions....the word shouldn't be there as it didn't exist at the time, and is, therefore, a misfit to the language of that era.

Tupael Khan-

Great work.

art H-

The Lord's Prayer I have always said debts and debtors, but I have heard it both ways. I use the New American Standard and I know for sure the KJ version said trespasses. I feel the same way about the Word of God being changed. The Word is so important to all believers, but when we trust in God, have Jesus Christ as our Savior and guided by the Holy Spirit in our hearts, we will be fine. You have helped me so much Stasha, and I truly thank you, this ME has been really hard on me. I found out about six days ago and have been really stressed over all this and after searching almost every video on YouTube in panic mode, and after I had prayed to God for help, I found you and your, videos which really calmed me down. You are a wonderful and beautiful person. Oh, I really believe just my opinion that CERN has been messing around and has gotten a hold of Dark Matter. That would be very bad.

Luke Berenstein-

This world is passing away, Jesus' return is imminent. Time is short for repentance and salvation, Jesus is calling

EVERYONE who can see these changes - the Holy Spirit brings all things to our remembrance - those who don't know Him yet, receive Him today, while there is still time... who can say if there will be a tomorrow anymore? Give glory to God, call on the name of the Lord Jesus Christ, and you SHALL be saved. Love and blessings to you all, my friends, and to your loved ones - I hope to meet you in the Kingdom.

Reboot-

Our Father, who art in heaven, hallowed be thy name. Thy kingdom come, thy will be done, on earth as it is in heaven. Give us this day our daily bread, and forgive us our trespasses, as we forgive those who trespass against us. And lead us not into temptation, but deliver us from evil. I haven't been to church in a while (and am now regretting it deeply), but know that for a fact. At first, I really thought that the Mandela Effect was possibly just a way that our minds interpreted the way we hear things into the way they should be spelled or stated logically (febreeze/febreze), but this is the icing on the cake. Out of everything in the church I have ever learned, the Lord's Prayer and the Nicene Creed were two things I knew by heart and never had to look at a book. The new Lord's Prayer is impossible. There is no way this can be real.

It is evident that some remarkable changes have occurred within the pages of the Holy Bible. But who administered these changes? Can we blame King James, Queen Elizabeth the 1st or the Council of Nicea for them? Or are these shifts a result of CERN opening portals that they should not. Has CERN opened a proverbial "portal to Hell" with this technology? Or is it something much *deeper? There is without a doubt a spiritual war happening on planet Earth; we can see it reflected all around us. No matter what is causing these changes, I still have the faith of God within my heart. Even if we woke up tomorrow to discover that all of the words in the Bible were gone, the real word of God is written on our hearts.*

We must reach deep within our souls and realize that if words change, our heart should not. I pray that with the publishing of this book, we will get answers to one of the greatest mysteries I have ever witnessed in human history. There are numerous Bible changes that I have been unable to cover here today, but I will continue to publish further books on this topic of Bible changes as the list of evidence is mounting daily. I feel confident that I have touched on the most significant collection of Mandela Effects to cross my path thus far in this book.

7. Theories

I can imagine that this book has been an eye-opening one for you to review. I, myself, am still flabbergasted by the Mandela Effects that are sent to me daily. My master lists of Mandela Effects has grown so extensive; I will be releasing further volumes of books shortly. One of the most important things that we can focus on now is what on earth is causing the Mandela Effect?

We have briefly reviewed a few of the possible theories throughout this book, but now I want to shift focus to some of the more fringe ideas brought to my attention throughout this journey. Just like the "Smoking Gun" chapter in this book, I also received my very first email regarding a Mandela Effect theory. The first contact that I received was from a person on YouTube calling themselves "Quantum Reality," I will share with you the email that he sent that eventually led me to unravel a never ending string of potential causes for the Mandela Effect:

> "Stasha, I have watched some other videos of yours exploring M.E. I have posted on a few of those. With your permission, I offer the following for your review. It is a little verbose but an easy read. I am going to proceed as if most will understand the concepts I am using We are familiar with Super-

string theory. We are told the math used to explain Super-string theory reveals a code that is identical to (not similar, identical) to the auto-correction code used in computer browsers. Now, we know how to sync a browser from one computer to another. All history, bookmarks, favorites, passwords, etc. can be imported into a browser on a different computer browser that had NOTHING in it. So, you turn off the first computer then turn on the second computer then sync the browser, and everything is there.

Remember I posted that Quantum Memories might be influencing reality? So, if we substitute computer browser settings with Quantum memories, it would be the memories that were transferred/imported. This would help explain why many, so things are different from what we remember, especially if more than one browser/reality was synced. This even explains why there are 7.5 billion people on this Earth opposed to the 6 billion on the Earth that I recall. This could even explain why even though we are no longer in the Sagittarius arm of the galaxy the constellations are similar along with the differences in Geography, History, Anatomy, etc.

This Shift may have been accomplished by an outside Entity / Multi-dimensional being. The Entity may not be a resident of Time and Space as we know it. (Outside of the Simulation) Using the Superstring and ' syncing the browser ' themes I mentioned before. We are told that at the sub-Quantum level that particles are comprised of Strings. These strings vibrate, with all matter having its' own frequency. Similar particles/elements would have their own frequency and corresponding harmonics. If you do a right click on a browser page and you can inspect the element (I use this feature in my job every day). All the code and much else can be viewed as how the page is constructed.

Using an interface (machine, apparatus, or whatever, we are talking super science here) to view the frequencies of matter in reality. Macros could be employed to select certain frequencies and corresponding harmonics. So, maybe those of us experiencing the Reality Shift Phenomenon have frequencies/harmonics in common with each other at the sub-Quantum level. For whatever reason, our frequencies were mapped and met the parameters of what frequencies to be: Saved, imported, synced and deployed in this Reality. Those involved in metaphysics have been suggesting something similar. I have tried to investigate this from a science angle. These may be converging theories. So, I think there actually may be something to the frequency posit.

-QuantumReality

I truly appreciate and can envision the theory presented to me by Quantum Reality. However, I will be the first to admit, that I do not know the first thing about Quantum Physics. Yet, string theory and quantum memories seem to arise and resonate with the Mandela Effect continuously. However, I can barely recite the periodic table of elements, so I have to accept the experts in these areas of study. Since I am in no way an expert in the field of Physics, I appreciate the research and ideas that come my way from trained individuals. I needed to connect with another expert to untangle this web of possibilities placed before me.

Since the most popular theories to cross my path have been directed in the categories above, I decided to reach out to another expert in Physics. I had to connect the dots of the Bible changes, plus the changes that may be caused by Scientists, themselves. In essence, I had to think outside of the box, not locate another "typical" Theoretical Physicist. The theory I am going to present to you was created by Byron Preston, AKA Harmony Mandela Effect. While browsing his YouTube channel during the process of this book, I located another series of videos, known as the "Throne of the Bible" series. I was blown away to discover that Byron also had a degree in Theology in addition to his background in Physics. His thorough research links the Holy Bible to physics. I now present to you "The Throne of God" AKA Physics of the Bible.

THE THRONE OF GOD
**Elucidates the Theory of Everything
And Fulfills Einstein's Unification Theory.
Now Explained by The Theory of HARMONY
By: Byron Preston, Copyright 1999.
All rights reserved to Author.**

Table of Contents:
Einstein's Unification Theory-
The Parables-
The Throne of God-
String Theory-

Supersymmetry & M-Theory-
The Implications-
Conclusion-
The Choice-

Einstein's Unification Theory-

Albert Einstein passed away with the notes to his Unification Theory on his desk at Princeton University. He was working on unifying all the forces of the Universe. Everything we can see and experience in this world can be explained by four forces; the Strong Nuclear Force, the Weak Nuclear Force, the Electromagnetic, and Gravity. Each of these forces is a special field of study in the world of physics. The problem is that each field has different theories that by themselves are perfectly sound. The problem arises when one tries to unify them. Working within the four-dimensional universe (height, width, and depth and time) science have not found an acceptable way to unify the four forces. Einstein, like most of us, concluded that these four forces must have an origin point. After all, the four forces can't just enter our universe out of nothing, and further, they can't just enter from four different directions originating out of nothing. The logical conclusion is that these four forces must have an origin point from within our universe.

Unfortunately, Einstein never could arrange these forces in a manner that would unify them and find an origin point. Let's take a look at the Bible for some expanded ways to view the universe. In the book of John chapter 14 verse 2 Jesus says, "In my Father's house are many mansions; if it were not so, I would have told you. I go to prepare a place for you". Theologians would interpret mansions as literally meaning, "dwelling places." They would further state that a better translation is "rooms," conveying the idea that there is ample space in heaven for all who come to Jesus as Savior.

Or could this be a hint at a deeper spiritual truth?

Let's look at the four-dimensional universe that we live in as a room. All the mathematical and physical equations represent furniture. For example, 2+2=4 would be a chair; Geometry is a couch, the laws of gravity is a dining room table, E=m c2 is a refrigerator, etc... Einstein and others insisted that all the laws of the universe, which are equations or furniture for our example, must fit in a three-dimensional universe, or room for our example. The problem is that there was, and still is, enough furniture for a mansion. So, Einstein and others went about trying to cram all this furniture for a mansion into a room. First, they try standing the couch up on its side against the wall, then put a coffee table on its side, the refrigerator in the corner with chairs on top, etc., until they run out of room. Insisting that it must all fit, they fry it again only with a different arrangement; a different order, only to fail every time. If you have ever put your furniture into storage, you know exactly what I am talking about.

The problem is that the furniture only looks natural when it is laid out in a house. You might be able to cram it all into a storage unit but it is not natural that way, nor is it functional. When it comes to cramming all the laws of physics into a tiny room not only would it be unnatural but the physicists could not even get it to all fit. It's easy to take furniture from just one room in a mansion and cram it into a studio apartment. It would be impossible to fit all of the furniture of a mansion into an storage unit the size of a studio apartment. It's simple to take just gravity and fit it into the three-dimensional universe, to only squeeze the laws of quantum physics into it. There just isn't enough room in a three-dimensional universe to fit all the laws of physics into it. Einstein, one of the greatest minds of all times, tried time and time again but, he couldn't ever get it to fit. In fact, Albert Einstein spent the last thirty years of his life trying to unify the four forces.

Einstein passing on made the theories he was working on available for others to examine and bring new ideas to the table. Of course, Einstein was right in concept; the problem was the parameters to his hypothesis. Once you knock down the parameters that he was trying to work within, a three-dimensional universe, unification of the forces could begin to take shape. Now there were enough rooms, in other words, a mansion. Instead of insisting that all of the laws of physics had to fit into a universe that we could see, touch, hear, smell, and taste, there could be other dimensions that we are unable to perceive. By having a hypothesis that there are no limitations, we began an era of new theoretical physics. This new field was called, "String Theory." It showed great promise at unifying the four forces and all the laws of physics. However, it ran into some problems and almost faded away, until it had a rebirth called "Supersymmetry." What's the difference you ask? That is something you can research on your own. For now, let's focus on the important issues.

String and Supersymmetry theories both prove in theory that the four forces and all the laws of physics begin to unify with a six string theory; in other words with the existence of six dimensions. The biggest problem was that there wasn't an origin point with a six-dimensional model. As the dimensions were increased and broadened, they found that all the furniture fit in a ten-dimensional universe, along with the elusive origin point. At last, scientists had their long sought after theory that unified all the four forces, laws of physics, and found the origin point of it all. For our example, they now had enough "rooms" to fit all the "furniture" of the "mansion" in. Problem?

Now that we have broken through the limitations of limited thinking and expanded the parameters of our universe to be ten-dimensional, why stop there? The theoretical physicist continued to expand the universe to 11,12,13,14,15,16,17,18,19,20, 21,22,23,23,24, and 25 dimensions. But, after ten dimensions the equations completely fell apart. Until the 26th dimension, all the equations came back together again. At this point, it appeared that there must be exactly 10 or 26 dimensions. Why stop there you ask? They didn't. They continued to expand the equations out to infinity. To no avail, the equations never come back together again. The conclusion was that there must be exactly 10 or 26 dimensions. The problem is which, and why isn't there more than 26? The second problem is that an enormous amount of energy is required to test this theory in experiments. This energy is not yet available, so we can't even, begin to experimentally prove or disprove this theory. This is where the debate of the greatest minds of our world begins. We are no longer looking for the "Unification Theory." Rather, we are looking for the "Theory of Everything."

Another quest is to determine the geometric shape of the universe. You are about to read how the Bible not only describes the moment before the beginning of creation, but it also gives the geometric shape of it, and one of the most advanced mathematical logarithms and physical constants in the universe. Using the Bible and this geometrical shape of the beginning of our universe, one can see how the universe was collapsed to create the event known as the "Big Bang." Knowing the geometrical shape of this moment, we can see how the universe expanded giving us the geometric shape of our universe today. Further, the book of Revelation chapter 4 in the Bible is really a description of the geometric layout of the universe today. Once this shape is established, it can be used as a 'computer' to confirm some of the most world-renowned theories of our day. More astounding it provides the means necessary to break through where all theories have fallen short; that is what my theory of "Harmony" is.

Albert Einstein said, "Raise new questions, explore new possibilities, regard old problems from a new angle."

In order to find the "Theory of Everything" we must explain it from the beginning; and further, before "the beginning." In fact, scientists for the longest time insisted that the universe was eternal, that there was neither a beginning nor an end to it. Einstein's General Theory of Relativity predicted that space was expanding and that all the matter in the universe is moving away from an apparent point of origin. Einstein, however, did not initially recognize this prediction. Astronomer William de Sitter found that Einstein had made a mathematical error. When corrected, de Sitter found the mathematical prediction that the universe was apparently expanding away from its point of origin. Einstein's theories provided the seed for numerous discoveries that point to a finite universe.

One of the most remarkable outcomes of Einstein's theories was the discovery that time itself is a physical property of the universe. In fact, it turns out that space and time are so tightly coupled to each other that one cannot exist without the other. Because of this coupling physicists now speak of "space-time."

Several years after Einstein published his equations on General Relativity; three British astrophysicists, Steven Hawking, George Ellis, and Roger Penrose turned their attention to the Theory of Relativity and its implications regarding our notions of time. In 1968 and 1970 they published papers in which they extended Einstein's Theory of General Relativity to include measurements of time and space. According to their calculations, time and space had a finite beginning that corresponded to the origin of matter and energy. Remarkably, they concluded that prior to that moment, space and time did not exist! Surprise, the Bible has already told us this.

Let's look at the beginning of the Bible, the Book of Genesis. "In the beginning, God created the heavens and the Earth. The Earth was without form and void, and darkness was on the face of the deep". This obviously means that there was a beginning before that space and time did not exist and that God created everything.

A deeper look is necessary. "In the beginning, God created," is the traditional translation of what is a somewhat complex and debated Hebrew sentence structure. Other translation possibilities have appeared in the last century, but they presuppose the existence of chaotic matter. Nothing in the remainder of Genesis, nor in the Bible as a whole, requires or necessarily recommends this view, even though such opinions are biblically tolerable. Still, the most direct and fully acceptable translation is the traditional one adopted here, "In the beginning God created." However, the existence of chaotic matter being biblically tolerable alludes to a deeper understanding of the universe.

The word void is defined in the dictionary as containing no matter, vacant. Clearly, this is what the Bible meant when they went on to say that it was without form. We have a mathematical logarithm that is a universal physical constant that describes the exact moment that matter becomes chaos. It is called the Mandella equation. Taken in reverse order, this simply could be a description of the exact moment before matter was created. Even more profound, the Bible has given us an exact description of what the universe was like before the creation of matter; it was chaos. Matter, space, or time did not exist. In comparison to the universe we know today, there was nothing. In conclusion, we have a universe that is described by the Mandella equation. However, we must keep in mind that this is only chaos compared to what we know as the universe of matter, that we live in today. We will get back to this, but for now, let's look at what Jesus has said about our present day universe.

I will begin with quoting Jesus describing Heaven in the King James Version of the Bible, St. Matthew chapter 13 verse 33:

> "Another parable spake he unto them; The kingdom of heaven is like unto leaven, which a woman took, and hid in three measures of meal, till the whole was leavened."

The proper way to interpret the Bible is to interpret it literally. When the literal interpretation is impossible, then we are to interpret it symbolically. Jesus used the word "like" in describing Heaven because there was not a conceivable way for people back then, (or even today for that matter), to fully comprehend Heaven. Unlike some other parables, Jesus never did explain this particular parable, leaving it open to interpretation. One such interpretation is that at present, the kingdom is not fully manifest, but at the consummation of the Age to Come, it will be known to all. Meanwhile, it does its work of permeating human society, penetrating evil and transforming lives. In the next chapter, we will look at some possible literal interpretations.

THE PARABLES:

I will now discuss what I believe the deep spiritual truth of the parable of the leaven and the mustard seed is in the book of Mathew. Each parable begins with "Another parable he..." which means that they are related to each other.

In Matthew 13:33, Jesus is referring to the three fields of physics. Contemporary physicists are only aware of two of them. They are the Quantum field and the field of General Relativity. These two areas explain all the laws of physics. However, they can't seem to unify the two fields. They are completely separate. That is because they are are missing a field of physics that has not been discovered yet, (or they simply do not know that they have already discovered it). It is a field of physics that is beyond the physical senses, the invisible field. Here, Jesus clearly tells us that there are three parts to Heaven. The leaven (which is like yeast used to make bread rise) is the energy that is placed and works in each field universe making all of life, as we know it, possible. The three measures of meal are the three fields of physics. The three containers protect each field universe, so it may rise and take form just as the meal (which is like bread dough) does, (see figure 1). So, the circle or container on the left is the field of General Relativity. The circle or container in the center is the invisible field. The circle or container on the right is the quantum field. The physicists were right that there are two fields that can not be unified with their understanding of the universe. Scientists will not be able to unify and find the origin of everything until one accepts that there is a field of physics that they can not detect with human senses. The Theory of Everything cannot be solved until everything is in the theory, (See figure 2).

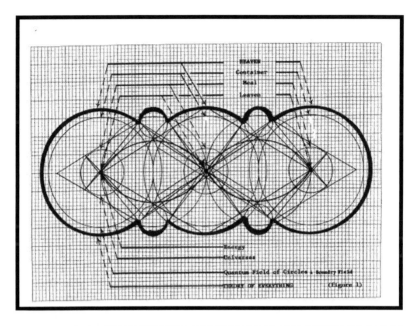

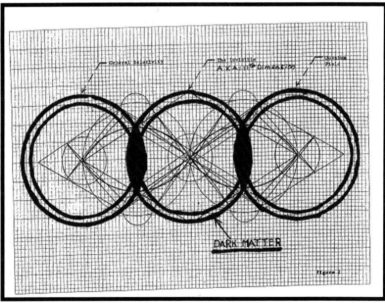

But, the woman hid the leaven. This represents that the field of physics and their origin are hidden and kept separate from each other. The General Relativity field of physics is kept distinct from the invisible center field of physics, and/or the outside field. The Quantum field on the outside is kept separate from the center field, and General Relativity.

"Which a woman took, and hid in three measures of meal, till the whole was leavened." When it is all leavened, simply represents, when everything is ready. When making bread you let it rise to a certain point, and then it is ready to bake. You wouldn't bake it before it has risen all the way, and if it rises too much, you not only don't bake it, but you have to push it back down and wait for it to rise again. She hid the three measures of meal until a moment in time. This moment is when the three measures of meal (3 fields of physics) are leavened (ready).

Then what? Perhaps, all will be revealed at this moment in time. We can find proof of such a time in the book of Mark chapter 4 verse 22 by Jesus telling us, "For there is nothing hidden which will not be revealed, nor has anything been kept secret but that it should come to light." But what is this moment in time for? What will happen?

Let's look at the Parable of the Mustard Seed in Matthew 13:31. "The kingdom of heaven is like a mustard seed, which a man took and sowed in his field, which indeed is the least of all the seeds; but when it is grown it is greater than the herbs and becomes a tree, so that the birds of the air come and nest in its branches." The theologian interpretation is that the parable of the mustard seed teaches the destined greatness of the kingdom. The kingdom fulfilled by Jesus now looks insignificant, but its greatness will be apparent in its consummation at the end of the Age.

In my theory of Harmony, the mustard seed represents physical matter. In comparison to all of God's creations, it may appear to be viewed as his least. The field is the universe. When it is grown, represents when it is ready, or leavened. The birds represent everyone that will be a part of the new heaven. They will nest in its branches, represents a new home for Gods kingdom. The book of Revelation chapter 21 says, "Now I saw a new heaven and a new earth.. ." It becomes clear that at a specific moment matter was created and it is transforming into something greater. We will discuss later as to what it is transforming into. But for now, how does the universe work?

The Throne of God:

The Apostle John wrote the book of Revelation between 70 and 95 AD. Chapter four is an attempt by the Apostle John to write down what he saw as he was in the Spirit and taken through a door in Heaven to see the "Throne of God." The Apostle John lived in a world that was not full of the technology that we have today. The people of that time even believed that the world was flat. Can you imagine trying to describe the awesomeness of the higher dimensions of the spirit world and the greatest task of all, to describe the Almighty's throne? This would be a difficult task for anyone today to accomplish. This is what John wrote:

Rev 4:3-8. And He who sat there was like a jasper and a sardius stone in appearance; and there was a rainbow around the throne, in appearance like an emerald.

4 Around the throne were twenty-four thrones, and on the thrones, I saw twenty-four elders sitting, clothed in white robes; and they had crowns of gold on their heads.

5 And from the throne proceeded lightnings, thunderings, and voices. Seven lamps of fire were burning before the throne, which are the seven Spirits of God.

6 Before the throne, there was a sea of glass, like crystal. And in the midst of the throne, and around the throne, were four living creatures full of eyes in front and in the back.

7 The first living creature was like a lion, the second living creature like a calf, the third living creature has a face like a man, and the fourth living creature was like a flying eagle.

8 The four living creatures, each having six wings, were full of eyes around and within. And they do not rest day or night, saying:

"Holy, holy, holy,

Lord God Almighty,

Who was and is and is to come!"

Let's first look at how the theologians would interpret this passage of the Bible. Twenty-four elders are the celestial representatives of all the redeemed, glorified and enthroned, who worship continuously. White robes symbolize purity. The crowns suggest victory and joy, not political authority. Lightnings and thunderings describe the awesome and wondrous power of God. Seven lamps of fire represent the seven Spirits of God, the Holy Spirit. Sea of glass, like crystal, denotes the unapproachableness and majesty of God. Four living creatures a cherubim, the highest-ranking celestial beings; they represent all the vital forces of creation whose primary function is worship. Full of eyes symbolizes unceasing watchfulness. The four symbols suggest majestic courage, strength, intelligence, and speed in the service of the Creator.

    The four living creatures were not alike in appearance. The theologians are correct when they say that the four beasts represent all the vital forces of creation. I seriously doubt that Paul knew, or even the theologians knew, that scientist would discover that four different and distinct forces can explain everything we see, touch, smell, hear and taste. These four forces can explain everything we experience. I believe that the four living creatures represent more than just cherubim, the highest-ranking celestial beings. The "Lion" represents the strong nuclear force. The "Calf" represents the weak nuclear force. The "Face like a Man" represents the electromagnetic force. The "Eagle" represents gravity. "And in the midst of the throne, and around the throne were four living creatures full of eyes in front and back." This represents the four forces physically around the thrown, and in the midst, (see figure 3). The eyes in front and in the back of the creature denote how the creatures/forces operate in the above paragraph.

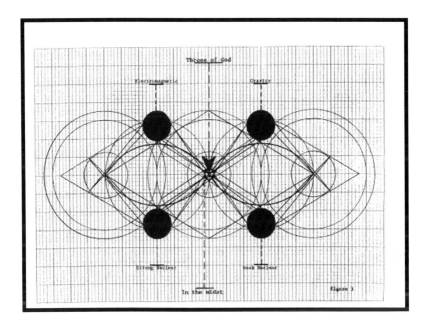

The colors in the center part of the painting starting from the top right clockwise are red, light blue, yellow, and dark blue. This is the rainbow that John sees around the throne of God. Keep in mind that he did not say that it was in front of or off to the side of the throne. The rainbow that he sees is the four living creatures in the midst of the throne. The twenty-four elders represent the twenty-four dimensions or strings that originate from the throne. (See figure 9) And count the lines that exit the circle.

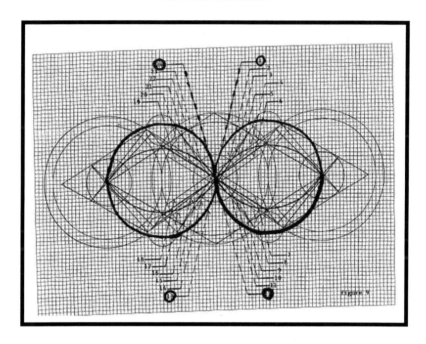

Keep in mind that the lines associated with the forces of the universe do not count as a dimension. The 25th dimension originates from the bottom, and the 26th-dimension originates from the top. Physicists agree that time and space are so connected that one could not exist without the other. Here the 25th dimension is space, and the 26th dimension is time, (see figure 14). Only the areas outside of the 25th and 26th dimensions are affected by it. This explains how the physical laws of physics and time do not bind the invisible field in the center. Not conforming to the reality that we know makes it difficult to accept and even further, to understand.

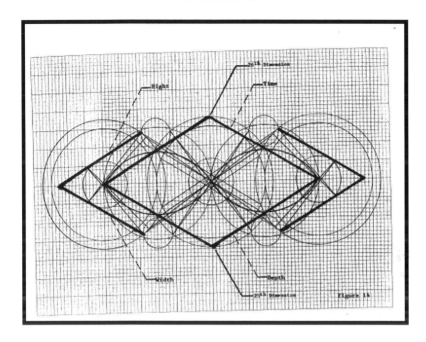

The center circle, field of physics, explains why string theories have only one model with 26 strings or dimensions. There is no matter and only forces in this model. (See figure 15) And count how many strings (lines) intersect our universe and the outer universe. Remarkable, the outer fields have ten each; 10 dimensions for the Quantum Field and ten dimensions for General Relativity. In this model, we have different parameters to the theory. They can agree on one 26-string theory but not on one 10string theory. Could this be because there is only one field of physics with 26 dimensions and no mass, and that there are two fields of physics that have ten dimensions, but differ in their makeup?

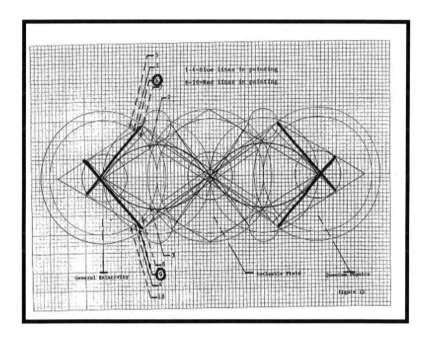

There are four different 10-string theories. Each theory is merely looking at various parts of the two outside fields of physics, (Quantum physics and General Relativity), leading to four separate 10-string theories, (see figure 4). In fact, this is the premise of the latest theory in theoretical physics. The theoretical physicists now say that all the string theories are correct, but they are merely looking at different "parts" of the "Theory of Everything." This new theory is called M-theory. Let's continue our interpretation of chapter 4 of the book of Revelation in the Bible.

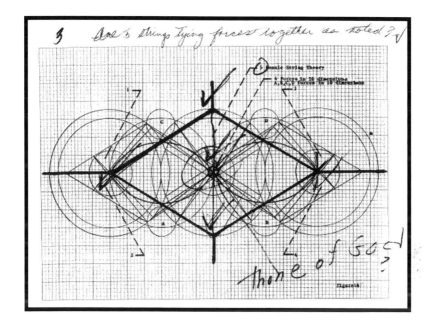

The lightnings and thunderings I will agree with the theologians' argument that this represents the wondrous power of God. But there is more to it. The four forces send their power straight toward the throne of God where they enter into their assigned "triangle" to have their force "focused" and turned back toward God, (see figure 5).

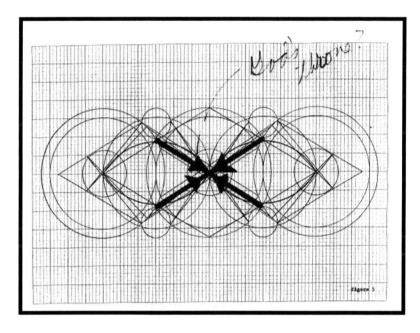

John describing the living creatures being full of eyes in the front and back depicts this. But here he is describing the four living creatures in the midst of the throne. They needed to see in front and in back to have the force from behind them pass through and to take in the force in front of them, focus it, and send the force back to God. The result of these actions is the lightnings that John saw. As all four forces reach God, he uses his awesome and wondrous powers to turn it into energy. This makes thundering sounds as God creates energy and guides this life force all over the universes, (see figure 6).

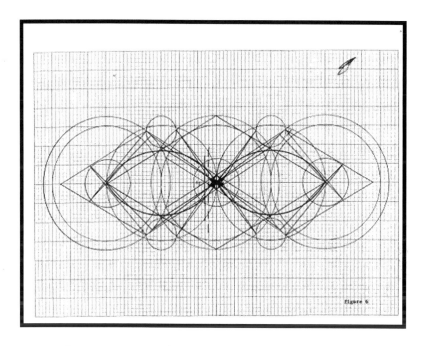

This new energy leaves the throne of God in an upward and outward direction. Used energy returns to the throne of God from the bottom up and inward where it is cleansed, (see figure 7) new energy is added, and God sends it back out into the universe (see figure 8). The energy leaving and returning to the throne of God is what John is describing as "the sea of glass."

The Mandela Effect

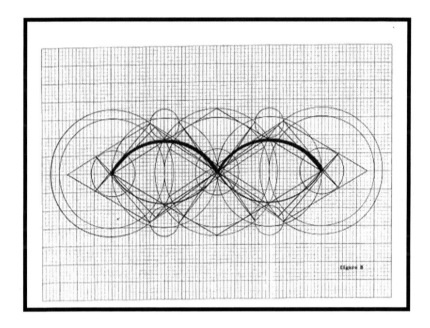

In reference to figure 9, I believe that all but two, of the dimensions, originate from the center circle of my painting, which is outside of the two 10-dimensional fields of physics, (see figure 15).

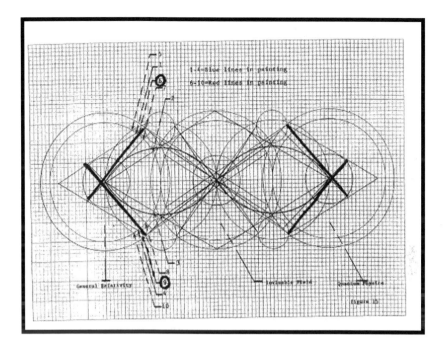

The center circle (the invisible field of physics) has four dimensions that the other two fields do not have. These four dimensions are used to stabilize the four forces of the universe, (see figure 10).

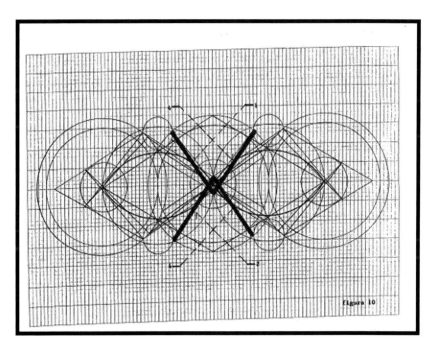

This brings us up to 24 dimensions total. Dimension 25 originates from the bottom center of the center circle, (see figure 11).

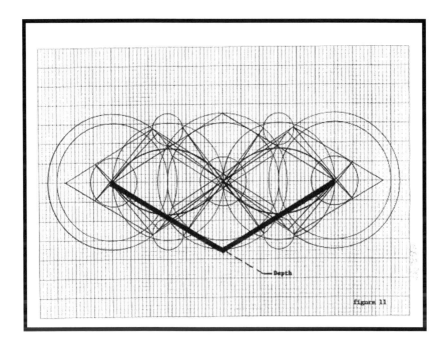

And dimension 26 originates from the top center of the center circle, (see figure 12).

Dimensions 25 & 26 originate outside of the 20 dimensions that are responsible for the laws of physics in the two outside fields, and the four dimensions that help to stabilize the four forces. In other words, they exist outside of the 24 dimensions that originate from the center field of physics. The four forces also have their home outside of the 24 dimensions. This makes it possible for the four forces to hold all three fields together, the "strings" in String Theory, Supersymmetry, and membranes in M-theory together. By dimensions 25 & 26 existing outside of the 24 dimensions and outside of the four forces that are in the midst of the center field, it makes it possible for them to effectively stabilize all three fields, 24 dimensions, and the four forces. The end result is a universe in HARMONY.

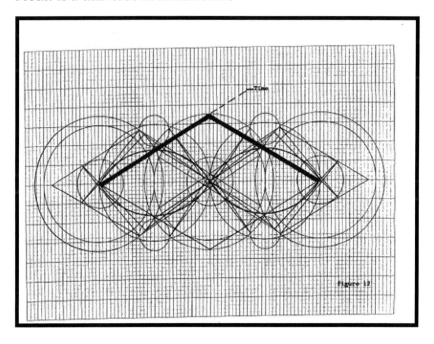

At this point, I have not found a theological interpretation of the meaning behind the living creatures having six wings each. Searching my painting for the meaning of the six wings each, you will notice that each of the four forces has three red lines going into it and three going out. These red lines are what John was symbolically describing as wings, (see figure 13).

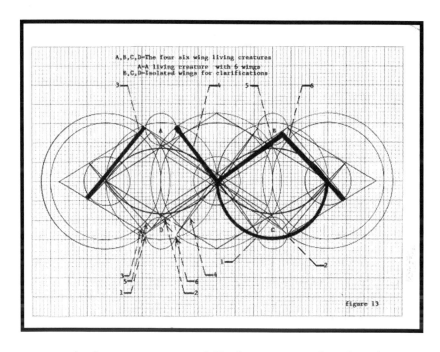

figure 13

The living creatures full of eyes around and within are two distinct descriptions of the creatures. This is because he is now describing their location in the universe, and their purpose here. Here he is describing how the forces act around the throne. We see that the four living creatures have a duel role. The eyes being around and within describes how they play a major role around the throne, which is the two outer fields of physics (General Relativity and Quantum physics), and within the center field of physics (the Invisible), the throne of God. But what was the meaning of them having six wings each? In string theory, the four forces begin to be explained with six strings or with a 6-dimensional universe.

In my theory of Harmony, each of the four forces has command of 6 dimensions each for a total of 24. This is why physicists believe that there are six dimensions curled up. This is why four different string theories have ten dimensions. It is because each theory is looking at one force with its six dimensions plus height, width, depth, and time, unaware of the center (invisible) field.

Scientists believe that at the moment of the Big Bang that dimensions 1 through 4 began to expand and dimensions 5 through 10 became increasingly smaller wound tightly up into a finite ball. My painting clearly depicts in the two outside fields that four lines are light blue and six lines are red. This explains why theoretical physicists have come to the conclusion that out of 10 dimensions there are two paths of evolution, divided into a group of four that are expanding and a group of six that are becoming increasingly smaller.

The "Seven Lamps of Fire" do represent the seven Spirits of God. This is the six triangles around the center circle and the diamond in the middle, (see figure 16). There are seven elements to God's creation. The four forces, the two outside physical fields of physics, and the invisible field of physics. It is only logical to assume that God's spirit must have a "hand" in each of these realms. The four triangles that look like a spike, (see figure 16), is God's Spirit working with the four forces. The two triangles to each side are the spirit of God operating the two outside fields. The diamond in the center is the spirit of God in the third field creating, directing, and overseeing all of his creations.

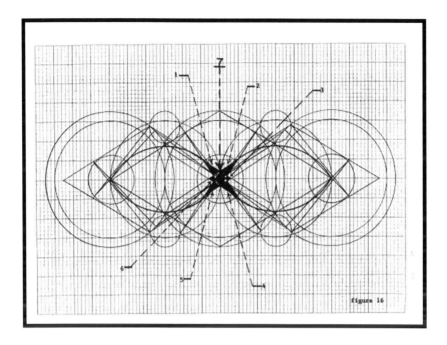

I will once again agree with the theologians that the sea of glass, like crystal, denotes the unapproachableness and majesty of God. In reality, the sea of glass is the whole energy of the four forces going toward the thrown and the Life Force energy that travels away from the thrown and out into the universe, (see Figure 5). God is the only one that can do this. It clearly denotes how his power is truly unapproachable and majestic.

In Revelation 4:11 The twenty-four elders worship God by saying, "You are worthy, O Lord, To receive glory and honor and power; For You created all things, And by Your will, they exist and were created." This clearly elucidates everything that I have described above.

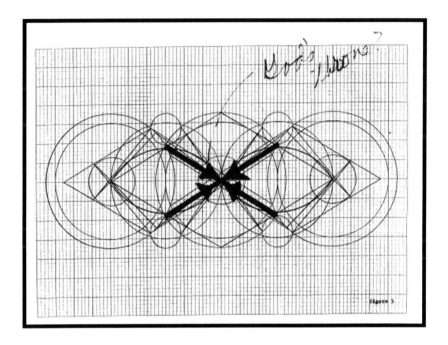

String Theory:

Let's take a look at the different string theories. There is only one theory that has twenty-six dimensions. It is called the Bosonic type. In this theory, there are only bosons, no fermions, which means that there are only forces, no matter, and both open and closed strings. The major flaw in this theory is that there must be the existence of a panicle called a Tachyon. A Tachyon is a hypothetical elementary particle that travels at speeds exceeding that of light.

According to the theory of Relativity, the speed of light is the limiting velocity for all ordinary material particles. Particles having nonzero rest mass can approach, but not reach, the speed of light, since their mass would become infinite at that speed. Theorists have argued that since nothing in principle prohibits the existence of a third class of particles that travel only at speeds exceeding that of light, such particles, may quite possibly exist.

Like the original theory of Relativity, the theory of Tachyons has several aspects that appear to contradict common sense but are fully self-consistent. For example, a tachyon must have an imaginary (in the mathematical sense) rest mass, or proper mass, and it must travel faster travel faster, rather than slow down when it loses energy. The center part of the painting is where the throne of God is. In this heavenly realm, matter, as we know it, does not exist; this is an agreement with the Bosonic 26 string theory. The center circle of the painting, (see figure 4), is a bosonic 26-dimensional field of physics. It is the only one of the three fields of physics that has all four forces fully present. All four forces are present around the center circle and in the midst of the center field, which is in agreement that there are only forces present in the 26-string theory. There are 26-strings that originate from the center field.

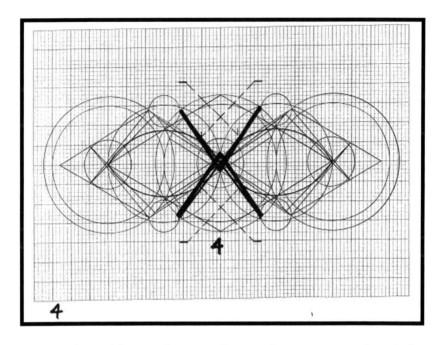

The Bible teaches us that God exists outside of the space-time dimension. A Tachyon being faster than the speed of light has escaped the space-time dimension and in fact, exists outside of it. I would deduce that this theory proves the existence of God, his angels, and his creations. Having beings that can travel faster than the speed of light, explains how our prayers can be heard in "real time" from our Father in Heaven. Existing outside the space-time dimension explains how God can and does predict events that are to come. It lets us know how God can be The Alpha and The Omega, Omnipotent and Omniscient. Going beyond the boundaries of common sense, but at the same time being sound, in theory, explains why God warned us not to try to understand Him, for we did not have the capability to fully comprehend and understand who He is in all of his glory.

Merely stating that God exists outside of space-time, that He is all-powerful and all-knowing, that there is neither a beginning nor end to Him, baffles most of us today. Although these statements made by God of who He is, as profound as they are, do pass the com-nonsense test. In fact, they must be true. Science with its 26-string theory has provided the means to prove the existence of God. It is no wonder that the Tachyon is what theoretical physicist considers to be the major flaw in this theory. It is not the major flaw; it is the answer that they are searching for.

I think it is appropriate at this time to quote astronomer Robert Jastrow, a professed agnostic:

"For the scientist who has lived by his faith in the power of reason, the story ends like a bad dream. He has scaled the mountains of ignorance; he is about to conquer the highest peak; as he pulls himself over the final rock, he is greeted by a band of theologians who have been sitting there for centuries."

When we look at 2 Peter 3:8 we find another example where theologians were ahead of scientists. It reads, "But, beloved do not forget this one thing, which with the Lord one day is a thousand years and a thousand years as one day." This is an aspect of Albert Einstein's theory of "Relativity'. One facet of this theory is that time is relative to where you are in the universe and how fast you are traveling. Light travels in waves. Time is measured by the frequency of the wavelengths that pass you. The faster you travel, the fewer times the wavelength of light passes you. Thus time is slower the faster one goes until they reach the speed of light. When one travels at the speed of light they are traveling as fast as the wavelength of light; therefore the wavelengths are no longer able to pass them. The result is that there is no unit to measure time. Therefore time stops at the speed of light.

Scientists thought of a way to test this theory. If it was correct, then you should be able to take two of the most accurate clocks in the world, atomic clocks, put one on the ground, and one on an airplane. Have the airplane fly around the world non-stop. When it gets back, because it was traveling faster than the clock on the ground, time should be slower for it. They did just that when the clock in the airplane returned it had lost 15 seconds. A Quartz clock loses about 15 seconds every two weeks. It is impossible for an atomic clock to lose 15 seconds in a 24-hour period. Unless the theory of "Relativity ' was correct.

This is only one facet of Albert Einstein's theory of "Relativity" that we were able to prove. There are parts of this theory dealing with gravity that we thought scientists would never be able to prove, but with the advancement of telescopes, we have been able to observe things in space that can only begin to be explained by Einstein's theories.

The faster you travel, the slower time is for you, until you reach the speed of light where time stops. Peter is clearly stating that the way you experience time is relative to where you are in the universe just as Albert Einstein said in his theory of "Relativity'. One day with the Lord in Heaven is the same as one thousand days on Earth and a thousand years on Earth is the same as one day with the Lord in Heaven.

It's astonishing how a person who lived in a time where people thought the world was flat and the universe revolved around the Earth knew about one facet of one of the most advanced theoretical physic theories of all time, that time is relative.

Supersymmetry & M-Theory:

What 10-string Supersymmetry theories have in common is that there is a Supersymmetry between forces and matter. Some have closed, and open strings and others have closed strings only. What is appealing to these theories is that you don't have the particle called the Tachyon. Of course, you don't, we have now moved to the two outer two fields of my painting where matter exists. The two outer fields have 10-string Supersymmetry each. If you were to remove the center field of my painting and push the outer two fields together, the strings (or lines) match up, (see figure 21). What drives this theory is matter. Since we can only perceive matter, it is only natural that scientist would want to eliminate the possibility of a field of physics, which does not contain matter, one that they can not measure. Once you eradicate the possibility of a field of physics that can not be measurable, one must push together the two physical fields of physics that do contain matter. Once this is done the strings appear to connect, thus, giving the appearance of a universe with only closed strings (see figure 21).

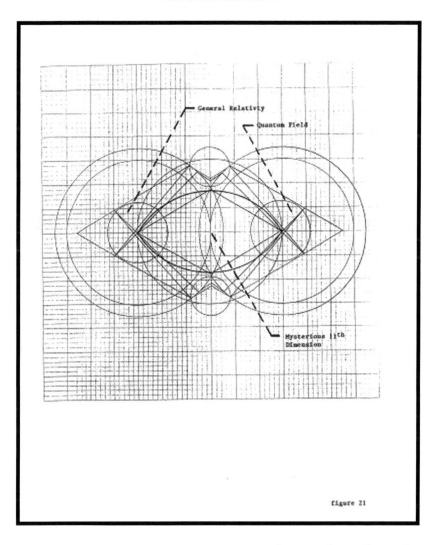

figure 21

Some 10-string Supersymmetry theories have the right moving and left moving strings differ. I won't go into a lot of detail about this. I will give you my explanation. Take two pencils. Wrap a rubber band around both of them with a twist, so you make a figure 8 with the rubber band. Now rotate both pencils outwards, and the rubber band will rotate smoothly. Now pretend that the center of figure 18 is the center point of the center field. This is the origin of all the strings. The pencils would represent the outer two fields. Two bands move outward and upward toward and around the two outer fields or pencils. After traveling through the two outer fields, and affecting everything in it, the rubber band returns toward to the origin or center field in an upward direction toward the center. The center field has no mass and no unit of spin. The two outer universes do contain mass and have two units of spin, outward and inward, or left and right moving.

The above paragraph describes 10-string Supersymmetry that requires a massless

graviton and two units of spin. It also describes the (xy=-yx) moment of strings. Ordinary space and time dimensions are described by ordinary numbers, which have the property that they commute: (xy=yx) the Supersymmetry directions are described by numbers that anti-commute: (xy=-yx). Where in (xy=yx) is an expression of the position of the universe, where (xy=-yx) is an expression of the movement of the universe? Both are correct in what they are describing. This is in agreement with my theory. As you spin the pencils outwards, you can feel how the rubber band from the center moves up and out. This is the +X and -X movement of strings. You can also feel how the rubber band from the center down moves up and inward. This is the +Y and -Y movement of strings, (see figure 18).

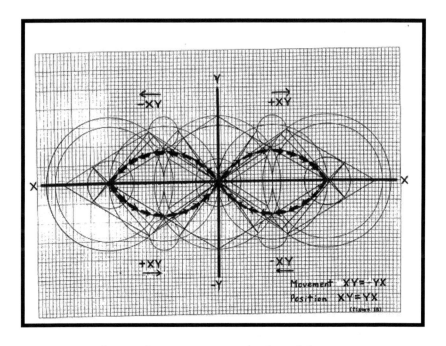

So, the only way to end the debate over a ten dimensional or 26-dimensional universe, a theory with no mass, and one that must contain mass, is to say that there are both. In theory, we can prove that there is one or the other, the problem is that one school of thought wants to argue with the other that they are right. The way we found that there must be 10 or 26 strings or dimensions was first to give up the idea that there are only three dimensions.

For example, figure 22 is Albert Einstein's universe. We have the four forces, time and space (height, width, and depth). Taking just one field of physics Einstein could find the origin point of the four forces as you can see in the center of figure 22. The problem is that this is just one field of physics. There was not enough room to unify both fields of physics, General Relativity, and Quantum physics. According to my theory of Harmony, this would require two separate fields of physics because they are separate but unified by a third undiscovered field of physics. He was working on the premise that our universe was bound by and only had three dimensions. Albert Einstein was trying to cram everything into one field of physics.

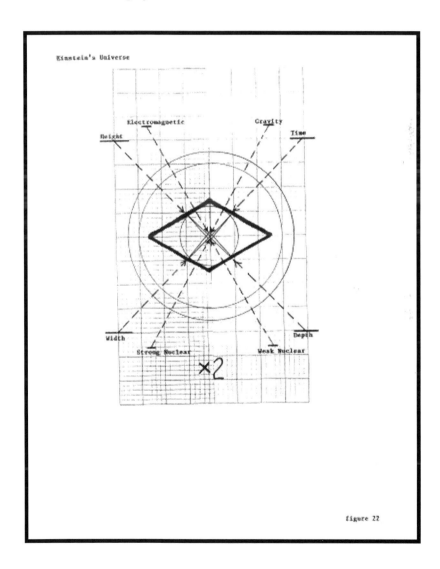

figure 22

The next generation of physics came along and said that there are ten dimensions. This opened up Einstein's universe to figure 21. Now we have the two separate fields of physic (general relativity and quantum physics), separate but at the same time unified by the extra dimensions describe as strings. The problem is that they did not reach their goal. The idea was that by unifying everything we could find the origin point. As you can see by looking at figure 21, there is no origin point.

Then M-theory came along. It said that all strings theories are correct because they are all just different parts of one larger theory, called M-theory, (see figure 4). But they don't like that 26-string theory. So we really need to look at figure 21. The Theoretical Physicists have now answered the question as to why everyone can have such different theories about the same thing and all be right at the same time. But, they still have not found the origin point. So, they are now saying that there must be an 11th dimension that exists outside of our ten-dimensional universe, (see figure 21). But, they stop there.

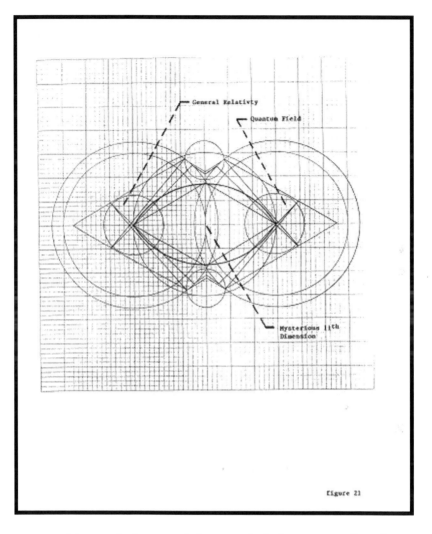

figure 21

This reminds me of the puzzle below. Draw four lines, without lifting your pencil, and connect all the dots. Go ahead and try it.

. . .
. . .
. . .

Unless you have done this before you can't do it. This is because the dots look like a box. You have been trained and conditioned your whole life to live inside a box. Well, have I got news for you. There is no box. It's an illusion! Look a figure 27 for the answer to this puzzle.

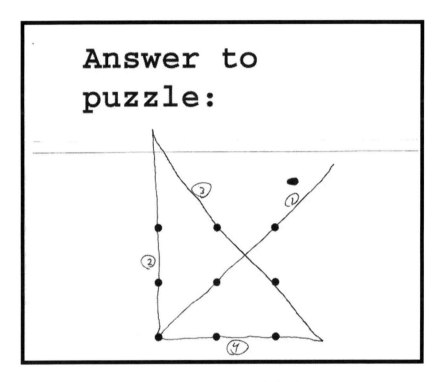

Now is the time to step out of the box one more time. To prove the Theory of Everything one must be willing to go beyond all the boundaries of the Human physical experience. To take a look from out there, looking back. Then and only then will one be able to say,

"AHHH, now I see."

My paintings and illustrations provide this view represented in a two-dimensional drawing. For example, a circle and a triangle on a piece of paper will never be equal; or will they? To a person living in a two-dimensional universe, they never will, and any such talk would be ridiculous and heresy. But in reality, one can lift the triangle out of the two-dimensional universe into the three-dimension universe and spin it, thus creating a cone. A cone sliced is not only a circle but, many circles and of different sizes. A triangle is many different size circles, a cone, and a triangle all at once and separately. This way of thinking is the only way to escape the boundaries of a physical 3-dimensional universe.

My theory of Harmony opens up the universe one more time. It exists far outside of the box. In fact, space-time or any of the laws of physics that we are bound by does not bind it. It exists outside of space-time. This is why the scientists are having such a toilsome time finding the origin point of our universe. That is why I call it the invisible field of physics. It can never be put into a lab to conduct experiments on to prove its' existence. We can only deduce that there is something missing in all of our theories that can only be explained by something that is beyond the human physical perception. Stepping outside the box and looking at my drawing one can see how this third circle, the third universe, the third field of physics is not only required to unify everything in existence but, it contains the origin that we all have been looking for.

There are exactly 10 and 26 dimensions. The two outside circles, (field of physics) have ten dimensions each. The center circle (field of physics) has 26 dimensions. And they are all correct when they say there are only four forces. In fact, the only thing we all can agree on is that there are only four forces. There is a reason for this. The Bible says that there are four living creatures, 24 four elders and that God and his angels are not bound by the same laws of physics that we are. This clearly is my painting of Harmony. To have more than four forces, you would have to add on a fourth large circle and two more forces to "hold it" all together. This would give us six forces and two invisible fields of physics. Well, the Holy Bible and our scientists even agree that this is not the case. There are only four living creatures and only four forces.

In Romans chapter 1 verse 20 it reads that, "For since the creation of the world His (GOD) invisible attributes are clearly seen, being understood by the things that are made, even His eternal power and Godhead, so that they are without excuse." The theologian interpretation is that in looking at the created world, every person should see the abundant evidence of God's existence and power.

The Implications:

What does all this mean? There will be a moment in time when all is revealed, and a new heaven will come about, but something must happen first. Let's turn to Matthew 13:49:

> *"So it will be at the end of the age. The angels will come forth, separate the wicked from among the just, and cast them into the furnace of fire. There will be wailing and gnashing of teeth."*
>
> *Why is it necessary to separate the wicked from among the just and cast them into outer darkness? The answer is Lucifer, who is now called Satan. The universe is not the same today as it used to be.*

Referring to the Mandella drawing, that is how the universe used to be. The book of Genesis 1:2 tells us that the "earth was without form and void."

The word void is defined in the dictionary as containing no matter. It is also described as chaos. The Mandella equation is one of the most advanced mathematical logarithms and physical constants in the universe, (see figure 23).

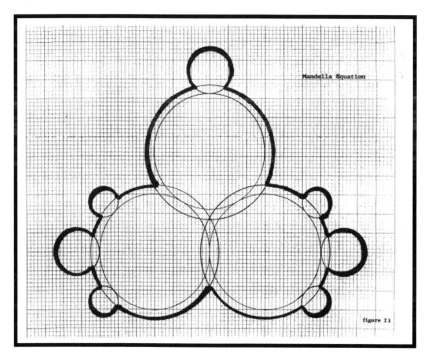

It is the exact moment that physical matter becomes chaos. Each of us is a being that exists in a physical universe. Anything else would be viewed as chaos to us. When in fact, as a spirit being, it is merely a different state of reality. There is nothing chaotic about it. In the Mandella equation, the four smaller outer circles at the lower half of the painting are the four forces of the universe, (see figure 19). In this universe, the four forces and all the "powers'" of the universe were available to everyone. In essences, anyone with the knowledge of how to use and control them could be "like" a god. Lucifer, being one of Gods highest, most powerful and closest angels, was aware of this and saw God use and control these forces. To him, it was evident that he could be "like" a god. He told his story to other angels and once enlightened with this knowledge they came to the erroneous conclusion that they could all be "like" gods. One-third of the angels sided with Lucifer and war broke out. Why did war break out? God and the two-thirds of the angels that remained loyal to God went about stopping Lucifer and those that sided with Lucifer.

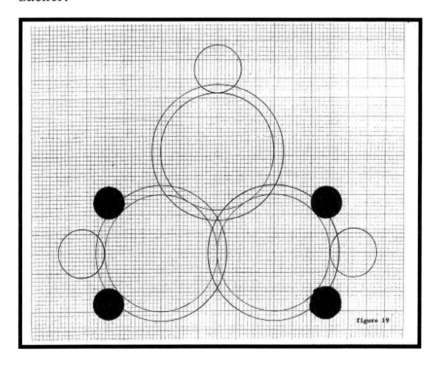

figure 19

God re-arranged the Universe to limit the power that the rebelling angels had and to restore peace to his Kingdom. Let's look at Chapter 12 verses 7-9 in the book of Revelation:

> *7 And war broke out in heaven: Michael and his angels fought against the dragon; and the dragon and his angels fought, eight but they did not prevail, nor was a place found for them in Heaven any longer.*
>
> *9 So the great dragon was cast out, that serpent of old called the Devil and Satan, who deceives the whole world; he was cast to the Earth, and his angels were cast out with him.*

The Bible says that One-third of the angels were cast out of heaven. This event is known as the fall of the angels. The word fall is defined in the dictionary as to collapse. The Mandella shaped universe was collapsed. Referring to figure 20 you can see how the universe collapsed. Nothing happens without a cause. Even if you want to say that "it just happened," something happened. Everything has a cause behind it; therefore there is the cause and a plan. A purpose, and design. The cosmological argument for God, as formulated by Aquinas and Al-Ghazali, asserts that everything that begins to exist must have a cause for its existence. Since the universe had a beginning to its existence (a fact seemingly verified by Twentieth-century science), then the universe must have a cause for its existence.

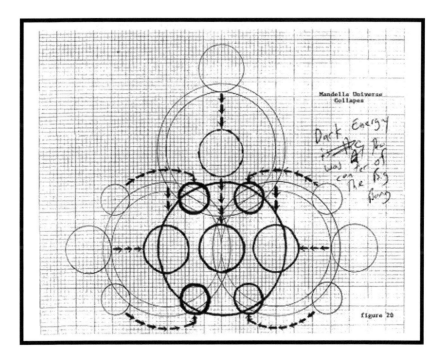

According to the "Big Bang" theory, the universe collapsed and then expanded. Scientists just don't know what the universe was like before the big bang or what caused it to collapse. They believe they know what happened less than a nanosecond after the Big Bang, but they just can't get down to the exact moment of creation, when time began, zero. Scientists say that the Big Bang is when time began. There is a reason why scientists cannot find a zero point in time. Referring to figure 25 the dashed lines are space-time.

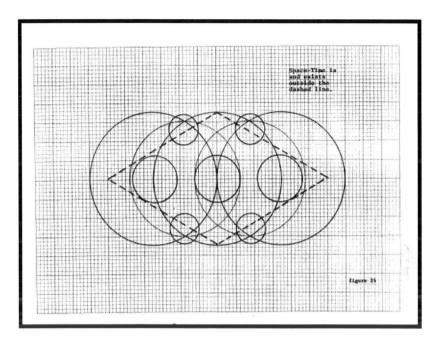

Space-time does not exist inside the dashed lines (diamond); it only exists outside the dashed lines (diamond). The Big Bang occurred inside the diamond, where time and space do not exist. The physical properties of space and time began for the two physical fields of physics once they moved outside the diamond.

The Mandella shaped universe was collapsed, (see figure 20).

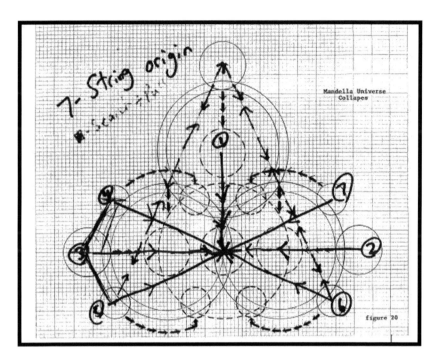

Causing the Big Bang to occur, (see figure 24). In an instant the two large outer circles were blown apart separating the two fields of physics with matter and fallen angels as part of its' makeup, leaving the third field of physics in the middle without matter, creating the universe as it is today, depicted by my theory of Harmony.

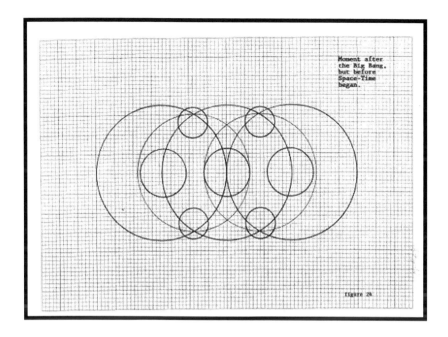

figure 28

Conclusion:

The Bible portrays the layout of the universe before time began, how it is today, and where it is destined. The universe consisted of a set of physical laws that are completely foreign to creatures of matter. The universe was re-arranged. Matter was blown into the two fields of physics that we are aware of today and the mysterious 11th dimension that M-Theory points to, which in fact, is the center part of my painting, the third, undiscovered and invisible field of physics.

Knowing the geometrical shape of the Big Bang, we saw how the universe expanded giving us the geometric shape of our universe today. Once this shape was established, I used it as a "computer" to confirm some of the most world-renowned theoretical physic theories of our day. More astounding, it provides the means necessary to break through where all theories have hit a wall.

That is what the theory of "Harmony" is.

Jesus tells us in the book of John 10: 16, "And other sheep I have which are not of this fold; them also I must bring, and they will hear My voice, and there will be one flock and one shepherd." The definition of the fold is a pen. The definition of a pen is a small fenced in area, to confine. The "sheep" being God's creations are separated from each other and are fenced in their field of physics. They are not in one field of physics but separate fields of physics that will be made into one field of physics with one shepherd. A universe in harmony. But Satan and his followers will not sanction a universe in harmony, therefore; they will be compelled to leave.

The Apostle John's description of the throne of God in chapter four in the book of Revelation gives us the schematics of how the universe is controlled and operated today. It is necessary to have a third field of physics to unify, balance, and control General Relativity and Quantum physics. The problem is that it will not stay this way. Looking at John 10:16, and Mathew 13:33 one can deduce that the universe is going to evolve once more. The next evolution will change reality once again. The laws of physics remaining constant are what maintains our universe, our reality. Change the Laws of Physics that maintain our universe, well, you change reality as we know it.

If the folds (separate field of physics) were opened up into one fold, then we would have one field (one universe) as we did before the creation of our present universe. If this occurred then, Satan and his followers would not be bound anymore. They would be free to cause havoc disturbing the harmony of the universe and to wage wars.

To make all three fields into one field, one fold; the disharmonic energy must be removed. The Bible states that the evil will be removed and cast into outer darkness, separated from the presence of God. The outer darkness is everything outside of the painting of Harmony; it is the only place that is separated from "the presence" of God.

In an effort to protect the greater good, God will be forced to remove Satan and his followers, casting them out. Thus, this evil will be cast into outer darkness, separated from "the presence" of God, where there will be wailing and gnashing of teeth. The disharmonic energy will no longer be able to be tolerated. This is why the Bible tells us that Satan and his followers will be cast into an outer darkness.

The more advanced we become in science, the more the scientist are concluding that there is a design to the universe, a finite beginning to it, and therefore there must be a creator. Some of the most professed agnostic scientists are coming to a conclusion that there is a God and becoming Christians. Those that believe by faith already know there is a God. Those that search for the truth through knowledge are forced to accept the fact that there is a design to the universe and therefore a designer. That designer is God.

Brothers and Sisters, it is time to realize that the truth of God, his plan, and his purpose can be found in the Holy Bible.

The Choice:

What it all comes down to now is, do You Want to be with God or separated from God? Do you want to be part of the new kingdom or cast into outer darkness?

It is very simple to be part of God's Kingdom. He is waiting with open arms for you.

1. Confess the sins that YOU know you are committing to God.

2. Turn away from your sins. Have a change of heart. Desire not to Sin anymore.

3. Confess that Jesus Christ is the Son of God.

4. Accept Jesus Christ as your personal Lord and Savior.

5. Receive a remission of your sins. Fully accept that God has forgiven you of your sins.

6. Jesus said that "In my father's house are many rooms: if it were not so, I would have told you. I go to prepare a place for you."

7. If you were to die today, do you know if you would be joining Jesus in the place that he has prepared for you?

How do I know that I will not be cast out?

What it all comes down to now is, do You Want to be with God or separated from God? Do you want to be part of the new kingdom or cast into outer darkness? It is very simple to be part of God's Kingdom. He is waiting with open arms for you.

1. Confess the sins that YOU know you are committing to God.

2. Turn away from your sins. Have a change of heart. Desire not to Sin anymore.

3. Confess that Jesus Christ is the Son of God.

4. Accept Jesus Christ as your personal Lord and Savior.

5. Receive a remission of your sins. Fully accept that God has forgiven you of your sins.

Jesus said that "In my father's house are many rooms: if it were not so, I would have told you. I go to prepare a place for you."

Original Diagrams from Byron Preston's Paper © 1999:

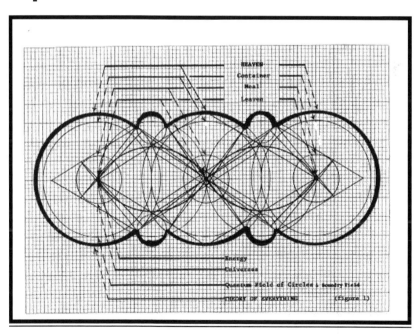

The Mandela Effect

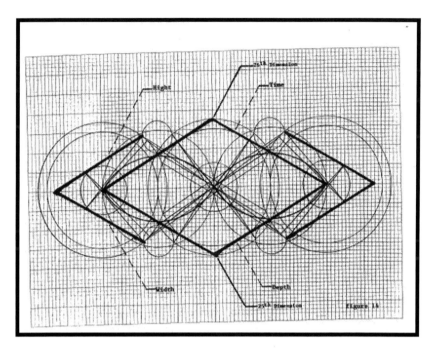

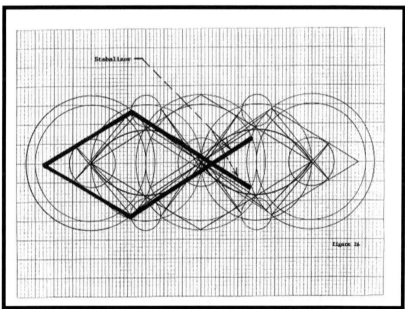

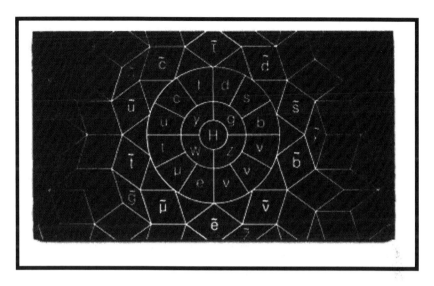

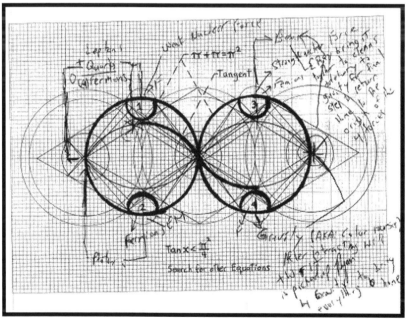

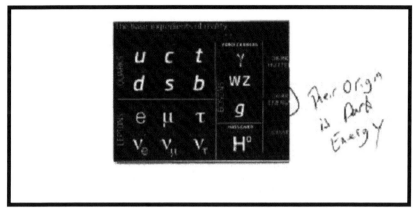

(Their Origin is Dark Energy).

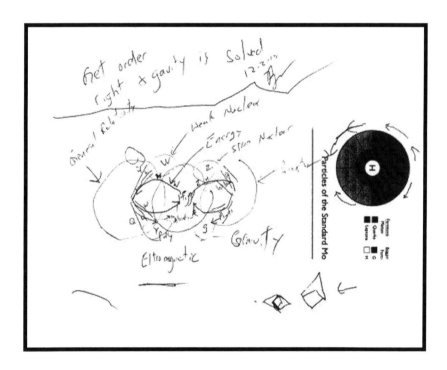

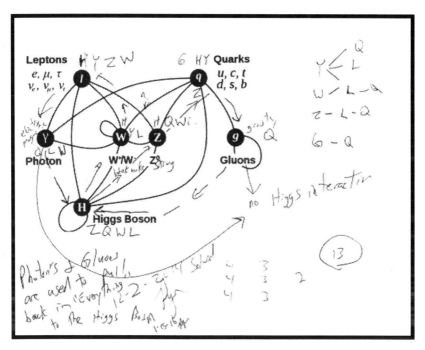

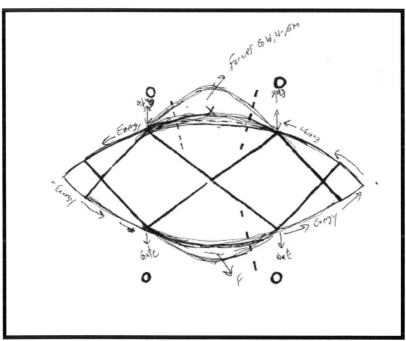

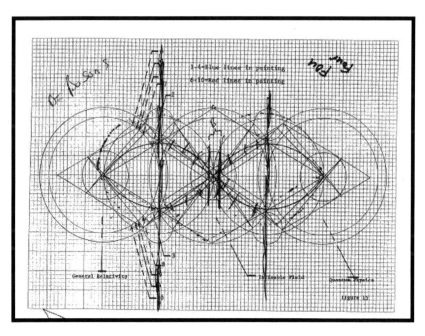

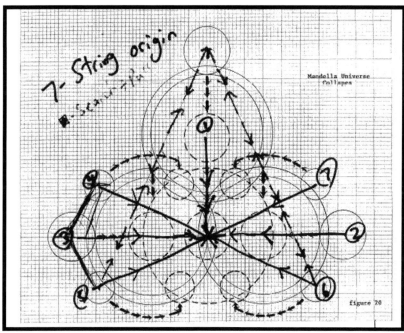

There are certainly some important topics that Byron has covered in this paper. I want to thank him wholeheartedly for allowing me to present his research to all of you. Byron had copywritten his diagrams and theory in 1999. However, he had yet to publish this work for the world to see. I am honored that he has allowed me to share this work with all of you finally. There are unbelievable connections to the Holy Bible and Physics; I feel this is just the tip of the iceberg. Do you resonate with Byron's Throne of God paper? Without having knowledge of Physics, this theory makes lots of sense to me. I hope that this can open your mind a bit more to the connections between God and Science. Coming into this project, I honestly did not think that scientists and theologians could find common ground, but Byron has proven me wrong. As he explains, God created everything we see; nothing is disconnected.

Following suit, I collected the remaining research that I could locate on the topics of Mandela Effect theories. I then recorded a YouTube video on the top 25 ideas that could be causing the Mandela Effect. I want to briefly summarize the ideas that came across my path while researching this controversial topic, before I could attempt to complete this first book. I have seriously just been capable of scratching the surface on the subject for all of you. I cannot wait to organize the remaining lists of Mandela Effects that are sitting in my desktop folder. But as I have discovered on this path, the changes never end, and if I do not find a stopping point, then this book would never get published. I will continue to connect dots until we have an answer to this phenomenon. But for now, we will continue with the contents of my Mandela Effect Theories video:

Mandela Effect Top 25 Theories- CERN, Time Travel, and The Great Deception:

https://youtu.be/sU2PTKc8Pak

In this video, I discuss the attacks and negativity that I had received since first embarking on my Mandela Effect video journey. I had no idea how many people I would upset for bringing attention to this effect. I began receiving comments and emails daily saying that I was crazy, a government agent, a witch, and scaring people with my content. I am always one who looks on the bright side of things, particularly when situations are difficult for people to understand, but I never saw the imminent danger of attacks coming my way. However, I did not, and *will* not, let this stop my research; this mission is a divine one for me. The closer I get to answers about this effect, the more attacks I receive. This is just a confirmation to me that I am on to something huge.

I continue to express my concern for the Mandela Effected, and my mission to complete the book that you are reading right now. I have been inspired to finish this controversial book for every single one of us who knows that the Mandela Effect is *real*. No one can tell us if our memories are "right" or "wrong" they may just be different from yours. I then go on to display the top 25 possible causes for the Mandela Effect. So I do not give away all of the ideas in one book, I will focus on the top 10 possible causes in this book and will conclude with the remaining theories in the second book:

1. **CERN:** We have discussed CERN at length throughout this book. The leading theory is that CERN has opened up a portal to another dimension while searching for the God particle, and since that day, our planet has merged with another consciousness and allowed other beings to enter our atmosphere through a rip in time in space. They have also been recorded to be experimenting with dark matter, which is extremely dangerous.

   I found a post on Reddit that expands upon this theory: "The LHC was turned on in early 2010, that's the same period when most people started reporting the Mandela effect start to happen. It's been said that the LHC has the capacity to destroy the universe if used improperly. It's also supposed that we live in a multiverse. What I propose is that the LHC IS destroying universes (hence the statue of Shiva destroyer statue or world's outside of CERN) and that every time a universe is destroyed our souls are merged with a universe that wasn't, and we carry on only from time to time noticing differences between our own personal memories and what is in recorded history."

   Retrieved from:
   https://www.reddit.com/r/MandelaEffect/comments/3566cp/cern_lhc_and_mandela_effect_connections_i_think/

2. **D-WAVE Quantum Computer:** The processor used in the D-Wave One code-named "Rainier," performs a single mathematical operation, discrete optimization. Rainier uses quantum annealing to solve optimization problems. The D-Wave One is claimed to be the world's first commercially available quantum computer system. Many people theorize that a quantum computer can actually be "alive" and giving commands and prompt to an A.I. that could control and change reality as we know it.

   "Quantum computers aren't new-fangled supercomputers. Quantum computers will allow us to access hidden dimensions in our universe that will give us more computing power than we could ever imagine possible. We didn't just make that up. That's taken directly from a speech provided by the other founder of D-Wave, Eric Ladizinsky."

   Retrieved from:
   http://www.nanalyze.com/2016/10/mandela-effect-quantum-computers/

3. **TIME TRAVEL:** Multiple people believe that the Mandela Effect could be a symptom of human beings time traveling. Similar to the movie "Back to the Future" if something was changed in the past, it would have a ripple effect that would alter the future as we know it. "Time travel. If it's real — and many physicists insist that time travel has to be, does it explain some of the Mandela Effect?

I'm reading The Yoga of Time Travel, by physicist Fred Alan Wolf. In the Introduction of that book, he reminds us, "... a scientific basis for time travel was established more than a hundred years ago... Albert Einstein and Hermann Minkowski showed how it was theoretically possible in 1905 and 1908." In the next paragraph, Wolf said something that startled me. It confirms something we've talked about here at MandelaEffect.com. "...let me tell you a secret: Some of the remarkable people you meet in life are time travelers.

A few of these people know it; the others time travel without realizing it, but they do it just the same. These are the people who appear older than their years or, yes, often enough considerably younger." So, how could this work with the Mandela Effect? As I see it, only for particular, limited memories. Here's an example. Let's say it's December 1986 and you're a teenager. You're aware of turmoil in South Africa, and — in your reality — Nelson Mandela is taking part in another hunger strike where he's imprisoned. (A side note for those who are looking at patterns related to 2s and 3s: He was prisoner 46664.) One morning, you go through your bedroom door and — in another reality where it's December 2013 (but you're only half-awake as you shovel in a quick breakfast, and you don't notice some odd changes) — your mother tells you the sad news that Nelson Mandela has died. Since your mind is on an upcoming exam, you assume Mandela died from the hunger strike. And then you go back to your bedroom, through the doorway* ...and you were back in 1986. That day's exam (at school) is a disaster. You know you did badly. You scramble to earn some extra credits *before school vacation because you need a good grade in that class. And, then it's the holiday season.*

You forget all about Mandela's death... until 2010 or so, when a friend says a few people recall Mandela dying in the late 20th century. Suddenly, that memory — which you think is from 1986 — comes flooding back. You know Mandela died in 1986, because your mother said so, and she was never, ever wrong about that kind of thing. Could that explain one kind of Mandela Effect? Maybe. I think it would only work for specific, private memories. And, you'd need to be oblivious to the cues that you're — at least briefly — in a different time. But, if time travel is as commonplace as Fred Alan Wolf suggests, it's something to consider."

Retrieved from: http://mandelaeffect.com/is-time-travel-part-of-the-mandela-effect/

4. **PARALLEL REALITIES:** The prevailing theory out there about this whole phenomenon is that it is a byproduct of the fact that there are multi-dimensions and alternative universes, and thus an infinite number of alternate realities. This is a theory that is also supported by modern quantum physics. Is the fact that there are some people who have alternative memories actual proof of parallel/alternative realities/universes? The Concept of 'Alternate Realities'? We are collectively becoming more and more aware that we all have multiple alternative realities happening at once. You can read more about

that here. In fact, the Angels recently told in a channeling me as follows: "Be the Divine creature that you are in all spheres of life, the afterlife, all directions of 'time and space,' because it is all happening instantaneously. All of your past, present and futures are going on all at once Dear Ones, and you will begin to understand this more like Quantum Physics and breaks through the 'surface' of reality and throws old perspectives askew."

Even though this sounds like the stuff of science fiction, those who are into the theories or are quite conscious of the 'reality' around them, really feel that it is real and happening. This has certainly been the case for me. Many people are starting to question reality and experience what modern quantum physics is suggesting - that we are all living in a holographic and simulated computer world of multi-dimensions. The theory is that the people that remember alternative facts had tapped into an alternative universe. So the alternate reality where Mandela died in the 80s no longer exists, or isn't the one that we are in right now. It is said that the more 'dominant' history takes over, the one that is stronger – and that eventually we forget or acclimatize - and it all becomes one through glitches in the system, although some still recall – and they are the ones questioning things. A Merging of Dimensions: Things are rapidly changing.

The reality is shifting before our eyes. Now, with the quickening of energies and energetic revolution on earth, some people might be moving in and out of or sliding between various alternative realities, and either not be conscious of it at all, or be somewhat mindful of small idiosyncrasies, glitches, oddities, or even a shift in the feel of the energy. Something feels different or out of place. As we move into different timelines, our consciousness is moving too. And with shifting timelines, history changes also. Our reality is changing before our eyes, including the written word, right down to how we spell things. Timelines and blurring, dimensions are merging. Some say that there are two earths – the old earth and the new earth, and they are converging, coalescing into one.

Retrieved from: http://www.nataliakuna.com/the-mandela-effect--alternate-universes.html

5. **POLE SHIFT:** The cataclysmic pole shift hypothesis suggests that there have been geologically rapid shifts in the relative positions of the modern-day geographic locations of the poles and the axis of rotation of the Earth, creating calamities such as floods and tectonic events. There is evidence of precession and changes in axial tilt, but this change is on much longer time-scales and does not involve relative motion of the spin axis on the planet. However, in what is known as true polar wander, the solid Earth can rotate on a fixed spin axis. Research shows that during the last 200 million years a total true polar wander of some 30° has occurred, but that no super-rapid shifts in the Earth's

pole were found during this period. A characteristic rate of true polar wander is 1° or less per million years.

Between approximately 790 and 810 million years ago, when the supercontinent Rodinia existed, two geologically rapid phases of true polar wander may have occurred. In each of these, the magnetic poles of the Earth shifted by approximately 55°. The geographic poles are defined by the points on the surface of the Earth that are intersected by the axis of rotation. The pole shift hypothesis describes a change in location of these poles on the underlying surface – a phenomenon distinct from the changes in axial orientation on the plane of the ecliptic that is caused by precession and nutation and is an amplified event of a true polar wander.

Pole shift hypotheses are not connected with plate tectonics, the well-accepted geological theory that the Earth's surface consists of solid plates which shift over a viscous, or semifluid asthenosphere; nor with continental drift, the corollary to plate tectonics which maintains that locations of the continents have moved slowly over the face of the Earth, resulting in the gradual emerging and breakup of continents and oceans over hundreds of millions of years. Pole shift hypotheses are not the same as a geomagnetic reversal, the periodic reversal of the Earth's magnetic field (effectively switching the north and south magnetic poles). If our poles have indeed shifted, then, of course, we would see changes in our maps and skies.

Not to mention, it could cause our North Pole ice cap to melt, as it has today.

Retrieved from:
https://en.wikipedia.org/wiki/Pole_shift_hypothesis

6. **NIBIRU or PLANET X:** Several theorists believe that we are now in the presence of an outside planetary influence, known as Planet X or Nibiru. If this foreign planet, sometimes referred to as "Planet 9" were to enter our atmosphere it would cause a shift in reality as we know it today. Also known as "Wormwood" in the Bible, The Nibiru cataclysm is a supposed disastrous encounter between the Earth and a large planetary object (either a collision or a near-miss) which certain groups believe will take place in the early 21st century. Believers in this doomsday event usually refer to this object as Planet X or Nibiru. The idea that a planet-sized object will collide with or closely pass by Earth shortly is not supported by any scientific evidence and has been rejected by astronomers and planetary scientists as pseudoscience and an Internet hoax.

   The idea was first put forward in 1995 by Nancy Lieder, founder of the website ZetaTalk. Lieder describes herself as a contactee with the ability to receive messages from extraterrestrials from the Zeta Reticuli star system through an implant in her brain. She states that she was chosen to warn mankind that the object would sweep through the inner Solar System in May

2003 (though that date was later postponed) causing Earth to undergo a physical pole shift that would destroy most of the humanity. The prediction has subsequently spread beyond Lieder's website and has been embraced by numerous Internet doomsday groups, most of which linked the event to the 2012 phenomenon.

Since 2012, the Nibiru cataclysm has frequently reappeared in the popular media; usually related to newsmaking astronomical objects such as Comet ISON or Planet Nine. Although the name "Nibiru" is derived from the works of the ancient astronaut writer Zecharia Sitchin and his interpretations of Babylonian and Sumerian mythology, he denied any connection between his work and various claims of a coming apocalypse. I tend to be one who looks on the bright side, and does not subscribe to the theory of "the apocalypse," but what do you believe?

Retrieved from:
https://en.wikipedia.org/wiki/Nibiru_cataclysm

7. **THE RAPTURE:** Some Mandela Effected individuals believe that the effect may be caused by what many people refer to as the rapture. In Christian eschatology, the rapture refers to the controversial "predicted" end time event when all Christian believers—living and resurrected dead—will rise into the sky and join Christ for eternity. Some Christians believe this event is predicted and described, without

the term "rapture," in Paul's First Epistle to the Thessalonians in the Bible 1 Thessalonians 4:17. The term "rapture" has come especially to distinguish this event from the event of the "Second Coming" of Jesus Christ to Earth, as some think is predicted elsewhere in the Bible, in Second Thessalonians, Gospel of Matthew, First Corinthians, and the Revelation. The term "rapture" is especially useful in discussing or disputing the exact timing or the scope of the event, particularly when asserting the "pre-tribulation" view that the rapture will occur before, not during, the Second Coming, with or without an extended Tribulation period. This is now the most common use of the term, especially among Christian theologians and fundamentalist Christians in the United States.

Other, earlier uses of "Rapture" were simply as a time for any mystical union with God or eternal life in Heaven with God. Catholics believe that the "Rapture" as a gathering with Christ in Heaven will take place, though they do not use the word "Rapture" to refer to this event, sometime during the second coming of Christ. There are many views among Christians regarding the timing of Christ's return (including whether it will occur in one event or two), and various aspects relating to the destination of the aerial gathering described in 1 Thessalonians 4.

Denominations such as Roman Catholics, Orthodox Christians, Lutherans, and Reformed Christians believe in a rapture only in the sense of a gathering with

Christ in Heaven after a general final resurrection, when Christ returns in his Second Coming. They do not believe that a group of people is left behind on earth for an extended Tribulation period after the events of 1 Thessalonians 4:17. So, do you believe that Jesus has possibly raptured all of the people with the Mandela Effect? You decide!

Retrieved from:
https://en.wikipedia.org/wiki/Rapture

8. **THE GREAT DECEPTION:** The Mandela Effect – Deception or Distraction? There has been a lot of talk recently about CERN and the Mandela Effect on the written Word of God ("logos"). It is suspected that we live in a multi-universe, and that CERN is destroying them. Their theory being that every time a universe is destroyed, our souls are merged with a universe that was not; we then carry on, only noticing from time to time, the differences between our personal memories and what is recorded in our individualized history. Along with many bizarre and strange things happening at CERN, there is much talk about something called the Mandela effect.

It became popular first in the paranormal/conspiracy world. The latest craze is centered upon people claiming that the Bible itself has been "supernaturally" altered, but how can any of this be true? I have been a long time Christian for over forty years plus, and none of our generation or before would even so much as taken this into consideration. The first thing recorded in the Bible in the Garden of Eden was when Satan disguised himself as a serpent and twisted the Word of God. We know the end of that matter. He did not twist it to the point of the obvious, but he did it with such cunning craftiness that it blew by Adam and Eve and the results were spiritual death and bondage. This plunged the entire world into Satan's control. Just think about that for a moment – he twisted one or two words and look at the hell he brought on this earth! People are demanding that words are mysteriously changing and/or being altered due to CERN's opening up of portals.

It is said that demons entering in through these portals, and through multi-dimensional universes, have been systematically changing the words written in the Bible. In addition, words said to be removed altogether are increasing in numbers, and people demand that this is reality. The fact of the matter is that this is not reality. When the first manuscripts were made, they were referred to as the master copies in the first century. The area was Antioch, Syria, which is known as the Northern Stream. This is where followers of Christ were first called Christians (Acts 11:19-26). This is also where Paul the Apostle first took his missionary journey. At this time, there was another version. It came out of Alexandria, Egypt. It is referred to as the Southern Stream. Alexandria was a port city, which was full of many nationalities and religions. Many Christians during the Great Persecutions traveled there. Origin, known as one of the earliest Church fathers, set up a school there known as Origin's School of Thought.

It was a contemporary school at the time, where the teachings were based on philosophy and espousing upon every written word in the Epistles saying, "Doth God really say _____?" It is by this method they altered the Word of God. This is where we get the Aryan doctrines that Jehovah's Witnesses use, as well as Mormon's and Catholics to name a few. To be able to get one's hands on the 'master copy' of the Northern Text was very difficult. The Christians from there guarded it with their very lives – literally. They knew the danger of allowing anyone at that time to have a valuable copy posed the risk of it being changed or altered. They knew the purity and preservation of the Lord's words were of tantamount importance. There was a process in receiving a copy. There had to be proof of identity of all men claiming to be missionaries. The master copies were hand scribed with perfect accuracy and seldom given away. After a time, Origin obtained a copy and he started his school. Since that time, there have been many versions of the Bible we now know of today.

The most accurate source, the Textus Receptus (Latin: "received text"), is the name given to the succession of the printed Greek texts of the New Testament, which constituted the translation base for the original German Lutheran Bible. This is the translation of the

New Testament into English by William Tyndale, known as the King James Version. Due to the growing number of believers through the centuries gaining ground all over the world, there were inaccuracies caused by translators who translated original texts into foreign languages generation after generation. However, it still stood as a base for translation that was the most accurate. I will not go any further in history. The bottom line is that God expected this to happen, and this is no surprise. Albeit, God preserves His own word and has done so since the beginning of time.

The Holy Spirit is the Witness Bearer that bears witness of the truth once spoken, and "Once delivered unto the Saints." There can be the most convincing arguments about these multi-universes in successions that are altering or altogether changing a word here or there, as they espouse, through layers of multi-universes coupled with human experiences that differ from one person to another. Here is a verse to springboard off: "In the beginning was the Word (personal pronoun), and the Word was with God, and the Word was God." John 1:1. "And He was clothed with a vesture dipped in blood: and His name is called The Word of God." Revelation 19:13.

Since we can all agree with the fact that the Word is God, and that His Word He speaks is not separate from Him, we then can also agree to the fact that God never changes. "So shall My Word be that goeth forth out of my mouth: it shall not return unto me void, but is shall accomplish that which I please, and it shall prosper in the thing whereto I sent it." – Isaiah 55:11. In other words, who can stop it! We know that our Father never changes. This means that if He is called "the Word of God", then He as God never changes. Even if the powers of hell, by help of human agency, were to alter words in the Bible through this deception, the end result will not matter – God still upholds His own Word. "He keepeth truth forever." Psalm 146:6. "Jesus Christ, the same yesterday, and today, and forever." Hebrews 13:8. "For I am the LORD, I change not...." Malachi 3:6 "Every good and perfect gift is from above, coming down from the Father of the heavenly lights, with whom there is no change of shifting shadow." James 1:17. "But you are the same, and Your years will not come to an end." Psalm 102:27. "Thy word O God is forever settled in heaven." Psalm 119:89.

God is the conservator of His own Word. Satan knows that the very Word of God is God. He knows he cannot do away with Him. The Lord knows that Satan works in the paranormal, and has the immense power to deceive, "if it were possible even the elect." He can perform paranormal acts to twist the Word, or use human engineering through the ages to remove God's words that were breathed into the Prophets by the inspiration of His Spirit, "into doctrines of demons." Lately, there are so many people who claim to be Christians arguing this truth about CERN, and the multi-universe warping of God's Word to one people group and then to another, that it is spreading like wildfire. This is Satan's power, and to accept that Shiva the Destroyer has the authority to remove and alter God's Word is a grave danger Remember these words of the Lord, "And whosoever shall speak a word against the Son of man, it shall be forgiven him: But unto him that blasphemes against the Holy Ghost it shall not be forgiven." Luke 12:10. These are harsh words. Nevertheless, it does say that all scripture is "God-breathed." 2 Timothy 3:16. Therefore, we deduce the fact that no matter how you slice it, if any words are altered there is a penalty for it. This is not to say you are to blame that you see God's words differently by the power of demonic suggestion, because the greatest enemy of God's Word is Satan.

God loves you; His word remains forever and never changes. The final result is that you place yourself into a deception that intends to do nothing more than to rob you. Time will be wasted, which is exactly what Satan is doing with this. This is a classic parlor trick of his. One last word: "But though we, or an angel from heaven, preach any other Gospel unto you then that we have preached unto you, let him be accursed." Galatians 1:8. So you see, though it is man altering it, or evil spirits promoting a supernatural dynamic, there are serious, eternal consequences. Do you believe that we are currently witnessing the great deception?

Retrieved from:
https://aminutetomidnite.com/blog/the-mandela-effect-deception-or-distraction/

9. **MIND CONTROL or HYPNOSIS:** The Case for 'Mind Control.'

   Some Christians believe that this is the sign of the devil, that we are being deceived and illusions are being created on a grand scale. Others believe that government agencies & secret power groups are mind controlling us. Steve Quayle from the Hagmann & Hagmann report claims that the Mandela effect is but a deliberate deception by those in authority and control.

   He states that this whole phenomenon is based on: "False memories which are an offshoot of the mind control MK Ultra experiments-alternate histories and shared memories which are being intentionally broadcast to change our perception of what's real and what's not a virtual "devil's mental playground." [....] The U.S. military and intelligence agencies are spending vast amounts of money to program the brain and to be able to both read your thoughts and control them and or implant them." Do you believe that the government could be messing with our minds?

   Retrieved from: http://www.nataliakuna.com/the-mandela-effect--alternate-universes.html

10. **PSYCHOLOGICAL OPERATION/ PSY-OP:** Psychological Operations (PsyOps) or Warfare "are planned operations to convey selected information and indicators to audiences to influence their emotions, motives, and objective reasoning, and ultimately the behavior of governments, organizations, groups, and individuals." There is a different level of PsyOps, but most conspiracy theorists concentrate on Black PsyOps. Or rather, ones that are hostile and denied by a government as having ever actually existed. Most of these programs are focused on opposing parties, but there are many instances where a government has experimented on their citizenry. One of the most notorious United States PsyOps, consisting of multiple subprojects/programs revolving around mind control — carried out by the Central Intelligence Agency (CIA) — is Project MKUltra.

As you can see, there's no questioning what lengths certain people and entities of power will go to when it comes to learning how to control and manipulate others. It's a scary thought. And MKUltra is only one of the known PsyOps declassified (and I'm not even sure if entirely). Imagine how many are still hidden from the public; both completed and currently playing out. Sometimes apologies have been forced, because of revelations, but don't think for a moment that the powers-that-be ever cease their diabolical plots!

Aliens, false flag operations, flat Earth, fake/holographic moon, time travel and Project Pegasus are just some of the subject matter believed to be falsely derived by governments to misinform and distract the masses. For this reason, people who aren't affected by the phenomenon (or have, but ignore and are in denial), clump The Mandela Effect in with the rest. You can readily find comments made on videos and articles by disbelievers calling The Mandela Effect a hoax and those attempting to raise awareness of their experiences government shills. Do you believe the Mandela Effect is a hoax?

Retrieved from:
https://crytonchronicles.com/ijak/2016/11/28/concerning-the-mandela-effect-part-seventeen

Now that we have reviewed the top ten possible causes of the Mandela Effect, I want to close out this section of theories by sharing with you the public comments I received on my Mandela Effect top twenty-five theories video. I think you will appreciate some of the comments I received, as they echo some of the ideas we have discussed here:

Pierce Bounce-

There was a mass extinction in 2012 the timeline had to be changed, some things on purpose others are domino effects (most examples of Mandela are domino effects). If everything were "changed back" your book wouldn't exist anymore because you wouldn't have had a basis to write it in the first place (if there was no Mandela effect), great channel by the way!

Rye S-

There was a guy with an icon of an American flag. I think his name was the USA, or it said American patriot or something like that I can't remember, but I have seen him say things to Mandella videos says "put the bong down" or "smoke another one." I feel like there's something to that. Trolls.

Tal S-

Remember "ChuckE Cheese," now it's "ChuckE Cheese's" that was always Chuck E Cheese for many decades...

damanofmen1958-

Thank you Stasha. Always love your insights. At least I know that I'm not crazy.

Stasha Eriksen-

We can be crazy together!

Life Matrix-

Agree on the zodiac shifts.

GiveToReceive-

The interesting thing I never in all four videos heard you talk about Svalbard archipelago. It did not exist before, and seed vault was on the continent. How do you remember it?

Stasha Eriksen-

When I moved here it was always here, I have mentioned it but just snuck into other videos. Svalbard is much bigger than my husband remembers, but the seed vault is new, it came maybe ten years ago, but yes its here!!

Unimatrix 001-

I have a video about we are on the other side of the Galaxy. Still the Milky Way. You see our new location in the Galaxy is now.

Glenn Green-

Love.

Leo Foster-

Learn to be psychic by decalcifying your pineal gland.

Lisa day-

I like your views and am 32. Things are happening that I can't explain. God is real. Thank you for your views.

Mari Adkins-

Have you watched any of Bluebeard's ME videos? If not, check him out.

Shawndell Rivers-

I believe you I don't think your crazy at all I have been Mandela effect maybe we are super human we are so much

more than a human beings just like you said on your part two video we are so much more than we think powerful beings with super power by shifted reality we created our parallel realities and when we dream too.

Clyde Robinson Jr-

An idea that I came into view with, The Brady Bunch.I think they put a "Frequency Signal" into the digital Broadcast of Television? I've asked people if they have stopped watching regular TV, most of us that are aware of this effect haven't watched Broadcast television in months. Now. That the signal isn't so relevant in our mind, we are now opening up to our new reality, Meaning that they knew about this phenomenon was coming for years. Then started broadcasting in a digital frequency to keep the masses at bay/ asleep in a matter of speaking??

Revelation "of St. John the Divine" is the way it's supposed to be for me. Because it was his vision of the events he was shown, and that is coming.

(Acts 19:1-5), what it means to "call upon the name of our Lord".. Then the "cross" spoken of isn't some artifact or shape people wear around their necks. But is the knowledge/ or truth of what Jesus said and taught, for it went against or was a cross of view held by the secular religions. For our Father nor his son Jesus Christ never had anything to do with the Religions of mankind. Thanks for your videos.

Mari Adkins-

Revelation "of St. John the Divine" -- same here.

angela tate-

Just because a person is a witch, it doesn't mean that Lucifer is their god, this is misinformation. Oh, and if you take the pentagram, for instance, it's a very ancient Christian symbol, only when it's inverted it can be used for nefarious purposes.

Rye S-

Christianity isn't "ancient." The pentagram is older than Christianity.

Stasha Eriksen-

I'm pretty sure the pentagram was turned into a cross, then back into a perfect cube. "Keeping you inside the box" but it is way older than Christianity. Satanists do not even believe in Satan, so there are a lot of misinformed people all over the globe!

PetLover2012-

You're a beautiful genuine soul; you share honestly and sincerely to help not harm. Thank you.<3

Ethereal Dublinerz-

Good medicine :)

Greg Gibson-

Mandela effect: fact not theory, if I were only 99% certain, I would NOT say anything!

I was aware of the Mandela effect for two months before my ribs spine and skull changed on June 13-2016 I had a floating rib on my front left lower rib cage, never had four short ribs only attached to my spine, and the Human spine was smooth,

The Human skull did not have a bone bridge under the temple !!!

My friend had a Human skull, and I would recommend stair at it every day for several years and in High school, I sat next to a skeleton in class, and I would play with it every day.

If You do not know what a floating rib is on Your front lower rib cage, then You are originally from here, A floating rib here is when one of the short back ribs are broken or detached from the spine, and it is painful, a floating rib where I came from is a short rib on the front lower rib cage attached with cartilage to the rib above it and is NOT painful every man has one on their front left lower rib cage and every woman has one on both,

There was an Old Testament story that explained why men had one less rib than women {God made Adam sleep and took one of Adam's ribs and formed Eve from Adam's rib} The stars here are different, the three stars of Orion's belt were straight across, the center star was 30% brighter than the left and the right star was the dimmest, the 3 Pyramids at Giza matched the size to brightness and the center Pyramid was the Great Pyramid and 30% larger, the small Pyramid was 50% larger, and they were north and south of each other, Here they are NE by SW of each other, Ecuador was directly south of the Gulf of California, the Panama Canal was east and west. Years ago I was told that there are 6 Planet Earths 5 of them were set up by the People that Humans evolve into, to ensure their existence from rogue or alien Time Travelers. "Time Wars" first they went back and recorded all of history, then they went back and collected stem cells from every Human that ever lived, so they could reproduce history, so if something happened they would have backups, and apparently, something happened to the Planet Earth I am from: (I did not want to believe this because it means that History is already written, more or less, CERN did not have a D-Wave: Quantum Computers were experimental.

What happened is the CERN on the Planet Earth I came from created a subatomic black hole or pulsar that grew and destroyed the Earth, and we were moved /evacuated here, by the People that Humans evolve into, they had swapped People from Earth to Earth throughout history to create continuity, the tiniest change in the timeline before a person was conceived will erase them, the chances of the same sperm cell and egg getting together is extremely low, so they had to hopscotch back in time collecting stem cells and implanting cloned embryos, clones are immune to timeline alterations, and pets are no exception, they are also cloned.

One of me on another Earth is named Jeff, they never swap anomalies, but they swap People that know them and I have met a few People that knew me as Jeff, and friends of the family that would call me Jeff once in a while, I found that strange as a child, People that did not know each other calling me Jeff and not having family or friends named Jeff.

The first Earth is the last Earth in a linear timeline, and the last Earth is the first, I was on the first Earth and from that perspective there were five other Earths in the past, a total of 6 Earths, Human genes are expressed differently on each Earth, on the first 3 Earths the only the difference is the ribs; the second Earth was almost identical to the first except men and women had the same one floating rib on each front lower rib cage, and the third Earth they Had three floating ribs on each front lower rib cage, I met a young lady in the early 90's that had ribs like that, we are currently on the fourth Earth, and we have four short back ribs. The conversion process is a constructed virus is used that produces specific proteins in the cells to turn on specific genes and turn off specific genes, to change how the Human genes are expressed, then they accelerate the person when they are asleep to a slightly higher dimensional state where metamorphosis takes place.

The virus is a constructed flu virus that infects every cell of the body the proteins it produces in the cells act as molecular keys locking and unlocking specific regions of the DNA changing how the genes are expressed. Each Earth has 5 Human gene expression modifier viruses one for each of the other Earths the gene expression of Animals is also modified from Earth to Earth, and they have Human to evolved Human gene expression modifier viruses, we already have all of the genes we need for all of our future evolution's, "They" told me they are us, we reunified are they. There are several things related to the Mandela effect that I did not cover here, such as the semantics of perceiving timeline alterations, merged consciousnesses I was not merged but some or most were, and the day I discovered the Mandela effect on YouTube I also discovered Nibiru / Planet X, and I remembered seeing on the news 3 weeks earlier that Greenland started its annual melt 1 month early, so I looked it up and Greenland here started its annual melt 2 months early, so I was brought here sometime between late March to early April 2016. I never heard the term "The Mandela Effect" before then, but I knew what it was. If you have any questions just ask.

HouseonHill-

Please research Nasa and realize they can't be trusted...please!

Stasha Eriksen-

Haha yea, that's pretty much what the video is all about. Never A Straight Answer! ;) = NASA

Tal S-

My theory, remember that tv series "Sliders," we may be in something like that. I use to doubt there was other realities & dimensions, but it seems like all that was hidden away is now being revealed. Things that were always a myth turns out myth is the opposite. A myth is true and fact. Those trolls, whoever they are, really need to grow up. You are very good at relaying the messages via YouTube to try to wake people up. I agree with everything you have been saying & have no fear.

Blair is a tank-

Hey Stasha I love your Mandela Effect videos, and I resonate with you a lot, and you just seem like a very smart person overall. This is my first life as I am a golden child and have never been incarnated before, I have to admit the government and elite have pissed me off in the past but I've realized over time that they've already been defeated so I don't let it bother me anymore. Blessings of love and light.

Dori Richard-

I found something for you! In 1973, an r45 record came out with the Lord's Prayer, and a Nun sang it. I looked up the video on YouTube, and she says trespasses! Here is a link: https://www.youtube.com/watch?v=Bd4iJkNCaZ8

Susso Anthony-

Hi Stasha,

I AM sorry about all the cruel comments others post ... don't even acknowledge them ...

WOW ... you Need to Sing before every video! Great Voice! There is no right and wrong!

The next right thing ...! Thoughts; feelings; knowing ... Constantly looking for Mandela Effects becomes meaningless without finding a meaning ... I thought that for a long time about schizophrenia - and with autism, they see people for who they are ... I wrote a large documentary on the life and death of Paul McCartney ... I haven't looked at it since the first shift that I experienced mid-summer last year. All of the audio mysteriously disappeared off of a USB and my phone ... Looking forward to us all getting to the bottom of this :)!

Love This Video ... Sharing! Looking forward to the book!

With Gratitude, Anthony

Life Matrix-

Yup the Brady's from another dimension. Still, have a lot of catching up on the Brady shows. Intrigued about the Beatles 411 you shared.

Ruben Santana-

Great info and video.

Ty very much for taking the time to do so.

Darrel Turner-

I am sorry you are getting so many negative comments, I hope the good ones outnumber the bad, i am a believer in sasquatch, and I have noticed on a great number of sasquatch pages the same thing with the trolls and haters, I would see the same lengthy comment on different pages,the commentator would have no subscriptions or any content,I see it so much, it seems like there is a considered effort from some organization, not just some unhappy people,,thank you for your work I know it takes time to film, edit and upload, it's appreciated, hope you have a great trip, if you spend any time in northern California, could you please say hi to the squatchers telepathically for me, THANK YOU -peace.

Paul-

Your Queen Elizabeth comments were gold. LOL

Fiorella Rainbowsky-

In 2013 I heard that timelines had shifted so maybe that could have caused this? :) I also read long ago in a book called Earth by the "Pleiadians" that eventually we would have completely different memories after timelines shifted so maybe this was their evil plan all along, and they didn't expect us to remember the truth. :(

Phil C-

Hi, great video, I didn't get to say what I thought was going on, there are several, one is that we are in constant flux, our spark of consciousness moves continually millisecond to millisecond through all the possibilities of matter and energy, but it seems constant. In Quantum physics that is called 'superposition' everything being everywhere at once, until we put our attention on it. As we do this, not only do we have a collective consensus but as the flux due to some big event, changes and the fluctuations become increasingly diverse from the one we were accustomed to, like ripples in the space and time continuum. You can see this in nature as one creature becomes another, cells converge and become all the stages of creation or evolution, or both at the same time. Creatures left the oceans and became amphibious; insects developed flight, reptiles and birds, mammals and so on all transformed. I think we experienced this before and to my way of thinking, we are doing so again. I think we have outgrown the present constraints; we would transform into a bird and fly. What will we become? Something more than we are, something better I hope. I think Cern and D-Wave computers and other devices have done something, but I believe they triggered a transformation that was coming anyway, maybe they dropped a bigger rock in the pond, and we all felt the waves!

There is one other that I remember from my old psychology days, and that is that we live in a world constructed in our heads and all our senses are interpretations of external information filtered through our world construct. In hypnosis, a man was hypnotized not to see his daughter who was standing right in front of him. The hypnotherapist then took a watch out of his pocket and placed it behind the man's daughter. He was then asked what the man was holding, he said a watch, he was asked what the inscription was on the watch, and he read that. This was behind his daughters back, and he should not have been able to see it. I had some experiences when I was a teenager, one was i was walking my dog at dusk along a path, and I saw in front of me an elderly black labrador walking towards me, I was just about to bend down and stroke it when it transformed into a sign saying keep off the grass, that was beside the path, not even on it. I felt stupid, but also intrigued as I was so certain it was an elderly black Labrador. What if everything is an interpretation in our heads? To look on the positive side it may be that we are now emerging from our self-imposed confinement in our heads, and now we are beginning to see the truth, or contrarily, is this just another illusion?

> *It is apparent that I am not alone in my experiences with the Mandela Effect and its many anomalies, I will never stop researching the cause. There is something much deeper happening in our universe than we truly understand. I want to thank all of my*

wonderful subscribers for their honesty and feedback. I also want to thank everyone who has supported me on this rollercoaster of an experience; I could not do this work without your love and respect.

Stasha Eriksen
www.stashaeriksen.com

8. Conclusion:

'What a long, strange trip it's been...'

Well folks, we have done it, we have touched on all of the anomalies that are linked to the Mandela Effect. We have researched the history behind the phenomenon, spoken with experts in the fields of Science, Physics, and Religion. We have discussed changes to our Human anatomy, shifts in our planetary systems, and most importantly, changes to our Holy Bibles.

We have listened to theories from all possible angles, CERN, D-WAVE Computers, the Rapture, the Great Deception, Time Travel, Parallel Realities, Hypnosis, Mind Control and even Psychological Operations.

We have included public comments and personal Mandela Effect stories from people of all ages, religions, geographic locations, and walks of life. I have done my best to include all points of view, even ones that debunked and contradicted the Mandela Effect itself. I have accomplished this work while trying to remain as unbiased as possible.

As a researcher, I never take anything as ultimate truth, until I have researched all of the facts, theories, and possibilities that could be causing the issue at hand. At the end of the day, I only want to get answers. I am heavily

Mandela Effected, and I know that if you are taking the time to read this book, that you are too…

Or at least, the curiosity is mounting?

No matter what is causing the Mandela Effect, one thing is for sure; Everything IS changing, and we are witnessing the shifts right in front of our very eyes. I am grateful every single day for having the Mandela Effect. This phenomenon has connected me with like-minded people all over planet Earth that I would have never had the pleasure of joining with had the M.E. not crossed my path.

I still do not know what is causing the effect, but I have my favorite theories. I will never stop researching this topic until we have the answers that we deserve. Something is going on here… that is undeniable, at best. I have focused this first book on what I consider to be the most important Mandela Effects. I have thousands of Mandela Effects that I have yet to present to all of you, so I will be releasing a second volume to this series in the very near future. If you have your own Mandela Effect stories, please send them to my email address, and I would be happy to include your story in the next book.

No matter what is causing the Mandela Effect, do not let it scare you, or instill unnecessary fear. Let the Mandela Effect be an eye-opening experience for you, one that keeps you questioning. Let this be the first day of your awakening.

Never forget your memories, and never give up on finding the truth!

In gratitude,

STASHA ERIKSEN

www.stashaeriksen.com
stashaeriksen@gmail.com

Made in the USA
Lexington, KY
07 September 2019